VENUS

Kosmos

A series exploring our expanding knowledge
of the cosmos through science and technology
and investigating historical, contemporary
and future developments as well as providing
guidance for all those interested in astronomy.

Series Editor: Peter Morris

Venus

William Sheehan and Sanjay Shridhar Limaye

REAKTION BOOKS

W. S.: *To David and Jane Sellers*
S. J.: *In memory of my parents and both sets of grandparents*

Published by Reaktion Books Ltd
Unit 32, Waterside
44–48 Wharf Road
London N1 7UX, UK
www.reaktionbooks.co.uk

First published 2022

Printed and bound in India by Replika Press Pvt.Ltd

A catalogue record for this book is available from the British Library

ISBN 978 1 78914 585 4

CONTENTS

PREFACE

After the Sun and Moon, Venus is the most brilliant natural object in the heavens, except for occasional bright supernovae or comets. It can be so striking at times that it may even cause alarm to casual stargazers. However, since it alternately swings to either side of the Sun between apparitions in the evening sky after sunset and in the morning sky before sunrise, there are periods as well in which it entirely disappears from view. It took careful observation to establish that the Evening Star and Morning Star were one and the same, which was first realized (as far as we know) by the Babylonians as early as the third millennium BCE.

With a telescope, the sight of Venus's phases as it orbits around the Sun was an important demonstration of the correctness of the Copernican theory over the Ptolemaic, and it still thrills most observers of the sky. However, apart from a view of the phase (the portion of the planet's illuminated disc that is visible), there are few enticements to the visual observer, since the planet's brilliant clouds prevent even a glimpse of the surface. With practice a few details can be consistently recognized through a small telescope in blue filter – such as the bright cloud swirls, which we now know are huge cyclonic storm systems, like our hurricanes, situated over the poles. There are also a few faint nebulous shadings which show up well only in ultraviolet (UV) images. The wavelengths of the spectrum they absorb are not readily identifiable with known chemical compounds, in addition to sulphur dioxide gas, and among the candidates to account for them, vast colonies of sulphur-metabolizing microorganisms cannot be ruled out. At the moment, the question remains one of many mysteries concerning our neighbour planet.

At the beginning of the spacecraft era, little was known about Venus – not even the rotation period – and speculation was rampant. Beneath the impenetrable clouds, the surface might consist of water oceans, jungles or deserts vacant but for sand dunes. The reality proves to be more interesting than we imagined. Venus's 'cloud-covered surface' stands atop what is by far the most massive atmosphere of any terrestrial planet, with a surface pressure ninety times greater than that of Earth. Moreover, that atmosphere consists almost entirely of carbon dioxide, and because of the now well-known 'greenhouse effect', being on the surface of Venus is like being inside a hot oven. Primordial heat in the interior (due to the decay of radioactive elements since the planet's formation) escapes only with difficulty, since the planet has no plate tectonics. The surface is volcanic, and huge lava flows have poured out and covered most of it (that is, everything older than perhaps a few hundred million years (Myr)). There are hundreds of shield volcanoes dotting the surface, and several lines of evidence suggest that volcanic eruptions are continuing up to the present time.

The planet is extremely inhospitable to life as we know it. And as much as we have learned about it in the last sixty years, the most interesting questions concerning Venus's evolutionary history are far from being resolved. Despite being almost the same age, size, mass and density as Earth, Venus has followed a very different evolutionary course. Earth and life have evolved together, even though we do not yet know how life began on our planet. What happened on Venus? We would desperately like to know, for learning the answers to such questions about Venus will also help us to better understand Earth and the Earth-sized exoplanets.

What role did Venus's proximity to the Sun have to play in its evolution? Did Venus retain much of its primordial atmosphere while Earth's was largely divested by a gigantic asteroid impact? Did it once have a moon? What happened to any early oceans it had – assuming it did have some, as is by no means certain – during the first few billion years (Gyr) of its history? How was its original, presumably rapid, rotation changed, so that it now rotates slowly and in the 'wrong' direction? Did Venus ever have plate tectonics? Did it ever have a 'carbonate-silicate' cycle that would have sequestered carbon dioxide in carbonate rocks, as on Earth, and kept carbon dioxide from escaping and being at large and contributing to the massive greenhouse effect? Did life form on Venus during the earlier part of its history?

A planet which once seemed (in our imaginations) so pleasantly and evocatively earthlike has proved to be far from it. What we have learned of Venus shows not only how precious Earth is, but how many different planetary destinies are possible, and how two planets, even of such similar outward characteristics, can end up looking completely different after a few Gyr of evolution. We can only imagine what variety there must be among the trillions of exosolar planets that we now know to exist in the universe at large.

Inhospitable as it may be for earthlings, Venus continues to fascinate, and remains a world of mystery.

ONE

THE 'EVENING STAR'

Bright Star! Would I were steadfast as thou art –
Not in lone splendour hung aloft the night
And watching, with eternal lids apart,
Like nature's patient sleepless Eremite,
The moving waters at their priestlike task
Of pure ablution round earth's human shores,
Or gazing on the new soft fallen mask
Of snow upon the mountains and the moors –
No – yet still steadfast, still unchangeable.

JOHN KEATS[1]

Photogenic Venus, approaching its greatest brightness, is shown passing 0.3° south of the famous open-cluster the Pleiades on 1 April 2020. The last time the two were close together in early April had been eight years previously, in 2012. This photograph was taken with a TMB 92-mm refractor.

A mong all the objects in the heavens, Venus is unmistakable, as it is far more brilliant than any other planet or star. It has attracted the attention of sky-watchers from prehistoric times, and the wonder it inspired is attested to in some of the oldest written documents we possess. Those who wrote them still communicate to us across a distance of some 5,000 years.

The motions of five planets, or 'wanderers' – Mercury, Venus, Mars, Jupiter and Saturn – were no doubt recognized before the dawn of recorded history. They fell into two groups: Mars, Jupiter and Saturn could travel all around the heavens, while Mercury and Venus were more confined, and remained within certain bounds of the Sun, either leading it into the morning sky or lagging behind in the evening sky.

We now know that the 'wanderers' are other worlds, in orbit (like Earth) round the Sun. Mercury and Venus lie closest to the Sun, and are known as 'inferior' planets; Mars, Jupiter and Saturn, as well as Uranus and Neptune, discovered only in the telescopic era (of course, in addition, there are other objects that are also planets – for instance, the dwarf planets Ceres and Pluto), lie further from the Sun, and are known as 'superior' planets. Venus is the planet which approaches closest to Earth. Moreover, being surrounded by a dense and highly reflective atmosphere of clouds, Venus outshines all the other objects in the night sky other than the Moon and the occasional supernova or a comet (of which the last observed in our galaxy was in 1604).

Not surprisingly, because of its brightness and the complicated pattern of its motions in the sky, Venus has been an object of intense interest from earliest times. It is named in the inscriptions of the first agricultural civilizations. These appeared in the Nile valley in Egypt, which runs from the cataract of Upper Egypt to the delta of Lower Egypt, where the annual flooding of the Nile produces fertile fields, and in the valley of the Tigris and Euphrates in Mesopotamia (the Greek word for 'land between the rivers').

Because the flooding of the Nile predictably follows a cycle, it was recognized by astronomical events. The flooding occurs in August at the time that Sirius rises just ahead of the Sun into the pre-dawn sky. This became the most important astronomical observation among the ancient Egyptians – their very livelihood seemed to depend upon it – and became the basis of their reckoning of the seasons. (Thus *akhet* was the season of inundation, *peret* that in which the land emerged from the flood and *shomu* that of the drought.) This led the Egyptians to adopt the first solar calendar in history, consisting of twelve months of thirty days each, and five additional days at the end of the year. The great historian of ancient astronomy Otto Neugebauer referred to this as 'the only intelligent calendar which ever existed in human history'.[2]

Though their calendar was intelligent and elegantly straightforward, the Egyptians had an extremely complicated religion, involving the worship of many gods. At an early stage, they worshipped the Sun as one of the chief gods, Ra.[3] The great river of heaven, the Ur-nes, marked the ecliptic, the Sun's path through the heavens, and along the river floated a boat whose passenger was a disc of fire, the Sun itself. The same stream carried the bark of the Moon (Iââhu; sometimes called the left eye of Horus), which appeared out of the 'door of the east' in the evening. The stream also carried the planets. Ûapshetatui (Jupiter), Kahiri (Saturn) and Sobkû (Mercury) steered their barks in a forward direction, like Ra and Iââhu, but Doshiri (Mars) sometimes oared backwards – which shows that the Egyptians must have been especially impressed by the length of the retrograde or backward loop the planet exhibits around the time it appears opposite the Sun in the sky. (Jupiter and Saturn also show such loops, but less prominently.)

In the Egyptian scheme, Venus received special treatment. It was regarded as a close confederate of the Sun, and had two names. It was Uati, the first star of the night, when it followed the Sun as an Evening Star, and Tiû-nûtiri, the harbinger of the Sun, when it preceded it as a Morning Star. It was also sometimes called Benin, the heron. This bird, still common along the banks of the Nile, dives under the river, then rises again, in the same way Venus, the celestial heron, disappears into the Underworld for long periods of time, but always returns.

The most assiduous Venus observers of antiquity – and, indeed, the most diligent at least until the Aztecs of Mesoamerica of circa 1000 CE – were the Sumer–Akkadians, the people of the other great agricultural civilization that developed in Mesopotamia (a term properly referring to the 'Fertile Crescent' between the Tigris and Euphrates rivers, north and northwest of Iraq in modern days, but by extension used to describe the region from the Zagros Mountains in the northeast to the Arabian Desert in the southwest).

Compared to the Egyptians, the Sumerians are little known. The Greek historian Herodotus, though he had much to say about Egypt, had never heard of Sumer, since it no longer existed in his lifetime. It was not until the middle of the nineteenth century – when then-assistant to the British ambassador in Constantinople Austen Henry Layard, with the assistance of Hormuzd Rassam, began excavating the site of Nineveh (near the modern city of Mosul, so badly damaged during the Iraq War of the early 2000s) that the recovery of its history and proud legacies began. Even so, it receives dismissive treatment in most histories. The Sumerians created, by means of a system of irrigation canals, the first urban society, introduced mass production techniques and evolved a system of writing based on cuneiform (a system of marks made in clay tablets with the slanted edge of a reed stylus) which at first was used mainly by merchants as a means to keep track of the flow of goods. As the Sumerian city-states became more powerful, the script was used to keep a large bureaucracy humming and to keep royal records in order. In time, the scribes began to keep track of not only the traffic of goods but the traffic in the heavens. Eventually they would note astronomical phenomena in records maintained for over a thousand years. These provided the basic data that nurtured the beginnings of astronomy.

Since the Sumerians, unlike the Egyptians, used a lunar calendar, they were keenly interested in determining when the thin crescent Moon could first be seen in the sky after sunset, an observation taken to mark the beginning of each new lunar month. To get a better view of the horizon, the sky-watchers began to observe from the elevated platforms of seven-level terraced ziggurats. The ziggurat of the ancient Sumerian city of Ur ('Ur of the Chaldees' in the Bible, the reputed birthplace of Abraham), built in the period 2112–2095 BCE, is the most famous. Among other tasks, the priest-astronomers were responsible for deciding when to add an extra, thirteenth, month to their lunar calendar in order to keep it synchronized with the seasons and religious festivals.

The crescent Moon (with 'Earthshine' illuminating the dark side), Mercury and Venus (just over the hill, above its reflection in the pond).

Venus and the Moon,
2 July 1921, as seen above
Camille Flammarion's
observatory at Juvisy, near
Paris.

Because of its brilliance in the night sky – sometimes it was even
known as the 'torch of heaven' – Venus attracted the attention of
the priest-astronomers, especially as it frequently lines up close to
the crescent Moon. Indeed, the sight of these two celestial bodies
together is among the most impressive phenomena of naked-eye
astronomy. Moreover, the Sumerian Venus, known as Inanna, was
the most important deity in the Sumerian pantheon. Because of this,
her wanderings through the heavens not only were of great interest
to the priest-astronomers, but inspired religious cults and gave rise
to a vast body of myth.

As seen from Earth, Venus passes alternately on the near
and the far sides of the Sun, and is then said to be at inferior and
superior conjunction, respectively. On Venus first becoming visible
to the naked eye in the evening sky (after superior conjunction),
the Sun lies some 6° below the horizon. Then, from evening to
evening, Venus makes a slow ascent, separating further and further
from the Sun while increasing in brightness and prominence. At
last, about two months after its first entrance on the stage of the
evening sky, it reaches a maximum separation (greatest elongation

east) of 46–47°, setting some four hours after the Sun. Now, it begins to drop sunward again, at first gradually and then at an ever-increasing pace. Some 144 days after it first appeared in the evening sky it is lost in the solar rays (at inferior conjunction), and is invisible for a time, typically a couple of weeks, before rapidly re-emerging into the morning sky as a Morning Star, west of the Sun. (The precise duration of its invisibility depends on a number of factors, including sky conditions, the observer's geographical latitude, his or her visual acuity, training, the difference between the azimuths of Venus and the setting or rising Sun, and so on.[4]) Now it holds forth as a Morning Star for 440 days before disappearing behind the Sun (at superior conjunction), after which the whole process repeats.

We know from a Sumerian text written sometime between 2150 and 1400 BCE called 'The Descent of Inanna' that she was originally a rain deity and fertility goddess. She was also bride of the god Dumuzi-Amaushumgalana, who represented the growth and fecundity of the date palm (hence she was sometimes known as the Lady of the Date Clusters). Setting her heart on ruling the Underworld and attempting to depose her sister (Ereshkigal, Lady of the Greater Earth), she failed in the attempt, was killed, and then dispatched to the Underworld. Eventually, Enki, the Lord of Sweet Waters in the Earth, managed to bring her back, but only on condition that she offer a substitute in her place. She chose her husband Dumuzi, when she found him feasting instead of mourning her absence. In the end, Dumuzi and his sister Geshtinanna were allowed to alternate as her substitute; each spent half a year in the Underworld, half a year above it. The myth of Inanna and her descent into the Underworld clearly reflects the planet's observed path in the sky.

As early as 3200 BCE Inanna was being worshipped in the Eanna temple in the city of Uruk, located about 250 km south of Baghdad on an ancient branch of the Euphrates called Warka (the biblical

Erech). The city was settled by about 4000 BCE and by 3200 BCE covered an area of at least 250 hectares (about 2.5 square km) and had 25,000–50,000 inhabitants.[5] Later, as one Sumerian city-state achieved dominance over another (thus Uruk was followed by Kish, then Nippur, Isin, Lagash, Eridu and Ur), her influence spread, and she acquired

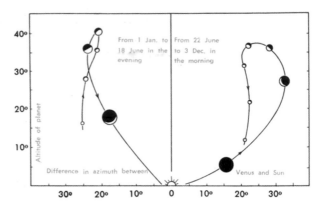

Venus's positions relative to the Sun in 1964. At left, from 1 January to 18 June, it appears as an Evening Star, then from 22 June to 3 December as a Morning Star.

the characteristic domains of the deities of the conquered city-states until she had achieved the status of being the most prominent female deity in ancient Mesopotamia. This may account for the fact that, 'unlike other gods, whose roles were static and domains limited, Inanna had a reputation for being young and impetuous, constantly striving for more power than she had been allotted', and notorious for her ill-treatment of her lovers.[6] In any case, in this way she acquired a dual nature as the goddess of both warfare and sexuality.

Her cult grew greatly after the conquest of the Sumerian city-states by the Semitic king Sargon I (the 'Great') of Akkad (c. 2350 BCE). Under Sargon, cuneiform not only maintained its record-keeping role but was adapted to phonetic writing capable of expressing the inflected Semitic language. Now, for the first time in history, spoken language could be rendered in detail, and this gave the Akkadians 'the ability to put in writing the host of texts that so far had been transmitted orally – not to mention the subsequent development of a high literary culture'.[7] It was one of Sargon's own daughters, Enheduanna, priestess of the goddess Inanna and the moon god Nanna (Sin) in the city of Ur, who contributed to the creation of this 'high literary culture' by producing several literary works, including the temple hymn 'The Exaltation of Inanna'; they deserve to be noticed not only because of their connection with Inanna but as the earliest literary works in history associated with a specific author's name.[8] Some idea of their quality can be gleaned from the opening lines (loosely translated by Michael R. Burch):

Lady of all divine powers,
Lady of the all-resplendent light,
Righteous Lady clothed in heavenly radiance,
Beloved Lady of An and Uraš,
Mistress of heaven with the holy diadem,
Who loves the beautiful headdress befitting the office of her
high priestess

. . .

Like a flood descending on alien lands, O Powerful One of
heaven and earth, you will teach them to fear Inanna![9]

In her dual roles as priestess of Inanna and of the Moon, Enheduanna would doubtless have taken special notice – as have sky-watchers in all times and places – of those occasions when Venus and the crescent Moon line up close together in the same field of observation. A particularly impressive rendering of such an event appears on a boundary stone of Meli-Shipak II, the 33rd king of the Kassite or third dynasty of Babylon (c. 1186–1172 BCE). Ishtar (Venus) is represented as an eight-pointed star, following behind the crescent Moon (Sin) and the Sun (Shamash), in a depiction which must record an actual observation. What is the significance of the eight-pointed star? It appears to be a clear reference to the celebrated eight-year 'Venus period', which was already known by the second millennium BCE.

As we now know, Venus takes 224.7 Earth days to orbit the Sun, travelling (as viewed from above the North Pole) in a counterclockwise direction. Earth, orbiting in the same direction, takes 365.26 days. As Venus passes from one point to the next in its motion – inferior conjunction, greatest elongation west of the Sun, superior conjunction, greatest elongation east – Earth falls behind

Venus. Venus, like a faster racehorse, catches up and passes Earth at each point. Finally, after 584 days have passed, the two planets return to the same relative positions to each other and the Sun (for example, inferior conjunction). The period of 584 days is important, and referred to as the synodic period (*synod* is Greek for 'meeting'). These positions are displaced at intervals around the Zodiac. After eight years, Earth has completed eight revolutions round the Sun; Venus, meanwhile, whose period of revolution round the Sun is 225 days, has completed almost exactly thirteen revolutions, so the whole cycle is reset and repeats.

If Earth were the centre of the motions of the planets, as some ancients believed, then the motions of Venus would resemble a lovely rosette-like figure technically known as an epitrochoid. This figure contains five lobes in which Venus approaches closest to Earth, and

Boundary stone of Meli-Shipak II, king of Babylon, showing Venus, the crescent Moon and the Sun in a line.

The 'pentagram of Venus', so called because it has five lobes where Venus makes its closest approach to Earth. Because of the near-coincidence in periods – in eight Earth years, Venus goes round the Sun almost thirteen times – wherever Venus is in the sky tonight it will be in the same place again in exactly eight years hence.

is sometimes referred to as the 'pentagram of Venus'. Each of the lobes is completed in a synodic period, so all five are completed in 5 × 584 = 2,920 days. But also, 8 × 365 days = 2,920 days. Thus the whole thing loops round perfectly, which means that wherever Venus is in the sky now it will be in the same place eight years from today. (The eight-year Venus-solar period is not quite exact; it falls short by about two and a half days on average, so that after eight years, the planet comes, say, to greatest elongation east of the Sun about two or three days earlier, and the times of these greatest elongations slowly drift back earlier and earlier in the season.) In India in the fifth century, Aryabhatta worked out the orbit of Venus round the Sun with a very good estimate of its period and size – but the observations he used are a mystery. It would seem to imply, however, that Hindu astronomy did not feature a geocentric universe.

The 'Venus tablet' of Ammisaduqa ('Enuma Anu Enlil' tablet no. 63, in the British Museum).

The Babylonian priest-astronomers certainly knew this eight-year Venus-solar period in Meli-Shipak II's time. In fact, they had already hit upon it several centuries earlier – by (at latest) the middle of the second millennium BCE. We know this from the famous 'Venus tablet', now in the British Museum. It once belonged to the library of the Assyrian king Ashurbanipal (c. 670–c. 630 BCE) and contains a record of the appearances of Venus – referred to as Ninsi'anna, the divine lady, illumination of heaven – during the 21-year reign of King Ammisaduqa, the penultimate king of the first Babylonian dynasty, which reached the meridian of its power under Hammurabi (r. 1792–1750 BCE).

The Babylonian priest-astronomers of Ammisaduqa's time seem to have had a much expanded workload, paying close attention not only to the heliacal risings and settings of Venus and the other planets (their first visibility before or after the Sun) but to eclipses of the Sun and Moon, the superior planets' stationary points, retrograde movements and changes in brightness; the planetary conjunctions with the Moon and stars and constellations, and with each other; and even their appearances within haloes about the Moon – for they were now no less interested in meteorological phenomena. They were keenly interested in everything going on in the sky, perhaps feeling something of what the art critic John Ruskin once expressed:

> The noblest scenes of the earth can be seen and known but by the few; it is not intended that man should live always in the midst of them . . . but the sky is for all . . . It is fitted in all its functions for the perpetual comfort and exalting of the heart, for soothing it and purifying it from its dross and dust.[10]

However, their observation was more like that of the herdsman or seafarer – they were interested in what the sky had to say about practical matters. Confounding correlation with cause and effect, they imagined that if an important event, such as the deposition of a king, an uprising, a famine, a war, followed upon some particular phenomenon they noticed in the sky, whenever the same sky phenomenon recurred the event would also follow. Thus the priest-astronomers wrote down on their tablets a vast quantity of data, including meteorological data and the planets' irregular motions among the stars, together with the terrestrial events presumed to follow upon them. This data provided the basis for the 'omens', the manifestations or interpreters of the gods, such as this one from the Venus tablet:

> In the month Abu on the sixth day Nin-dar-anna appears in the east; rains will be in the heavens, there will be devastations. Until the tenth day of Nisannu she stands in the east; at the eleventh day she disappears. Three months she stays away from the heavens; on the eleventh day of Duzu Nin-dar-anna flares up in the west. Hostility will be in the land; the crops will prosper.[11]

Notably, only the appearances of Venus are ominous, not the disappearances. This would seem to imply that the Babylonians of that time believed that precise positions of Venus's appearances on the horizon correlated with such life-and-death matters as cycles of rain and drought, of planting and harvest. Ammisaduqa and the other Babylonian kings of that era clearly consulted the priest-astronomers to find out what Venus had to say about the best times for making peace or waging war. Ironically, Ammisaduqa's own reign ended with a devastating raid from the neighbouring Hittites; within only twenty years, control of the city would pass to a new dynasty in the long and complicated history of the region.

Venus after Babylon

Though no people were more assiduous Venus-watchers than the Sumer–Akkadians and Babylonians, Inanna's cult, from its beginnings in Sumer, continued to find an enthusiastic following and she wielded enormous influence for millennia. According to the great Sumerian scholar Samuel Noah Kramer, the ancient Sumerian story of Inanna's descent into and return from the nether world – which is really the planet Venus's story – inspired 'Babylonian, Assyrian, Phoenician and biblical traditions (the last giving rise to Mohammedanism and Christianity), as well as serving as an important influence in the religions of the pagan Celts, Greeks, Romans, Slavs, and Germans'.[12] Hindu mythology has a different origin of Venus. Sukracharya ascended to the sky in the afterlife, having been a teacher of Asuras. He possessed the 'sanjeevani vidya', knowledge of bringing the dead back to life (perhaps tied to the disappearance and reappearance of Venus in the sky after solar conjunction). The origin is not usually recognized, but the influence of that story in the resurrection myths of today is still very much with us.

In Assyria and Babylon Inanna was worshipped as Ishtar, the 'Queen of Heaven', and her symbol was the eight-pointed star, or sometimes a lion. She was the same as the Ashtart or Ashtoreth worshipped in the ancient Levant among the Canaanites and Phoenicians, and known in Greek as Astarte. (Curiously, in Sanskrit the number eight is ashta.) The English poet John Milton would recall her as:

> Ashtoreth, whom the Phoenicians called
> Astarte, Queen of Heav'n, with crescent Horns;
> To whose bright Image nightly by the Moon
> Sidonian virgins paid their vows, and songs.[13]

Astarte's influence is felt in the Greek identification of the planet with Aphrodite, their goddess of love and beauty. In China, it was Tai-pe, 'the Beautiful White One'. Though most cultures have identified it with female beauty, this has not been universal; in India, it was Shukra, a masculine deity, and it also is identified as a male in some Native American stories.

The Homeric Greeks used the names Hesperos and Phosphoros, the first for the Evening Star, the second for the Morning Star. Thus in the *Iliad*, the glint of Achilles' spear is compared to 'Hesperos, the most beautiful star set in the sky'.[14] The later poet Sappho (c. 630–570 BCE) evokes more homely emotions:

You are the herdsman of evening

Hesperos, you herd
homeward whatever
Dawn's light dispersed:

You herd sheep – herd
goats – herd children
home to their mothers.[15]

The recognition that the Evening Star and the Morning Star are the same body came very late to Greece. The difficulty is no doubt related to the asymmetry in Venus's behaviour: in the 144 days or so it takes to pass between evening and morning elongation, Venus is on the near side to Earth of its orbit round the Sun, but in the 440 days it takes to pass between morning and evening elongation, it is on the far side, and so has to travel through a much greater distance. (Note that 144 + 440 = 584 days, the synodic period.) Unless one has reason to keep it under continuous observation for very long periods, as the Sumer–Akkadians and Babylonians had done over many centuries – and as the Maya of the Yucatán were to do later – it is far from trivial to grasp the secret of its peregrinations, or that Hesperos and Phosphoros are, as Tennyson wrote, a 'double name/ For what is one . . ./ Thy place is changed; thou art the same.'[16]

We are told that the famous Pythagoras of Samos was the first Greek to realize this, at the end of the sixth century BCE.[17] Pythagoras, himself a half-real, half-mythical person, who combined in equal measure rational thought and mystical obscurantism, is famous (or infamous) to every mathematics student because of his theorem regarding right-angled triangles. Along with Thales, Anaximander, Heraclitus and others from the Ionian islands, Pythagoras was one of what Aristotle called *physiologoi*, 'those who discoursed on nature'.[18] Since Pythagoras in his earlier years visited Egypt and Babylon (and very likely India, since a solution for the quadratic equation was known there), he probably learned of the Hesperos–Phosphoros identity during his extensive travels. However, he was anything but a derivative thinker. He made many original discoveries, most significantly the hidden numerical relationships in music – for instance, that pressing down on a fret in the middle produces an octave, while dividing it in the ratio 3:2 produces the musical interval known as the 'fifth'. This led him to propound the idea that there were hidden numerical secrets everywhere in Nature, hence his saying, 'All is number.' Either Pythagoras himself or one of his early followers attempted to apply this manner of thinking to the universe as a whole, devising a cosmology in which Venus and the other planets followed circular paths whose positions followed a musical progression.[19]

For the first time, the strange periodicities in the motions of Venus that had come to the attention of the assiduous Sumer–Akkadians and Babylonians (and would later be discovered anew by the Maya of Mesoamerica) could be understood not just as cycles but in terms of a cosmic geometry. Venus was on its way to becoming a world known to science.

TWO

THE TELESCOPE: A NEW PHASE BEGINS

Of light by far the greater part he took,
Transplanted from her cloudy shrine; and plac'd
In the Sun's orb, made porous to receive
And drink the liquid light; firm to retain
Her gather'd beams, great palace now of light.
Hither, as to their fountain, other stars
Repairing, in their golden urns draw light,
And hence the morning-planet gilds her horns;
By tincture or reflection they augment
Their small peculiar, though from human sight
So far remote, with diminution seen.

JOHN MILTON, *PARADISE LOST*, VII, 359–69

The ancient Greeks held that Earth was the centre of the universe, and that the Sun and other planets, including Venus, revolved round it. (An exception was Aristarchus of Samos, who in the third century BCE advanced a heliocentric theory in which Earth was a planet revolving around the Sun, but he found few followers at the time. Another was Aryabhattya in the fifth century CE, who worked out planetary orbits, but his contributions are largely unknown in the West.) The most elaborate form of the geocentric theory was set forth by Claudius Ptolemy, an astronomer who lived at Alexandria, Egypt, in the second century CE, in his *Almagest*.

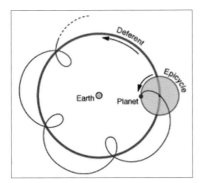

The motion of Venus according to Ptolemy. The motion round the large circle – the deferent – is one year (to keep Venus in lockstep with the Sun) and that round the epicycle 225 days. The combination of these motions produces the Venus pentagram, in which Venus makes five swings inwards to Earth on its epicycle in eight years.

This was the most influential text on astronomy until the early Renaissance. According to this theory, each planet moved round in a small circle known as an epicycle, whose centre in turn moved round in a larger circle known as the deferent. In the sixteenth century, a competing theory (still with epicycles, but now centred on the Sun rather than Earth) was worked out by the Polish astronomer Nicholas Copernicus (in Polish, Mikołaj Kopernik), a Catholic canon in Frombork cathedral. Copernicus drafted a sketch of his new theory as early as 1512, but cautiously bided his time before publishing, not least because he was afraid of being charged with heresy; he did not bring out the final version until 1543. This was *De revolutionibus orbium coelestium* (On the Revolutions of the Celestial Orbs), of which the first copies are said to have arrived when he was on his deathbed.[1]

The gauntlet had been thrown down, and later astronomers would set out their stalls in favour of either the Copernican or the Ptolemaic system. There was one particular observation, involving Venus, which could at least in principle decisively favour one over

Venus, moving round its epicycle as the centre of the epicycle moves on the deferent circle round Earth. A feature of the Ptolemaic system was that the centre of the epicycle always had to remain on a line between Earth and the Sun. Evidently Venus could appear new or as a thin crescent but never full, gibbous or even half.

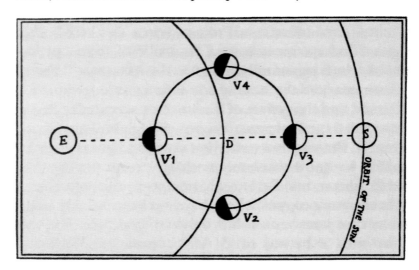

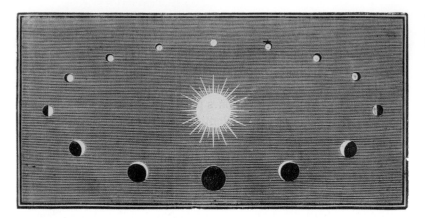

The phases of Venus, as modelled by the Copernican theory.

the other. In the Ptolemaic system, Venus always lies between Earth and the Sun, and its motion is as shown in the figure here. In the Copernican system, Venus travels all the way round the Sun, and so can appear both between Earth and the Sun, and on the far side of the Sun.

Consequently, assuming Venus shines by reflected sunlight (and even in Copernicus's time not everyone conceded this, some believing that Venus might be self-luminous or transparent), it would on Ptolemy's theory always show the crescent phase as it travelled round Earth. It could never be seen as full, or even half. But in the Copernican system, it ought to exhibit the whole range of phases – crescent, half, gibbous, and even a fully illuminated disc when on the far side of the Sun.

To make the crucial observation, optical aid is needed. In 1609–10, Galileo Galilei, a professor of mathematics at Padua and a follower of Copernicus, was inspired by the Dutch lens-maker Hans Lippershey's invention of what Galileo called a *perspicillum* and the German astronomer Johannes Kepler a telescope. As is well known, he used it to discover the craters and mountains of the Moon, and the four moons revolving about Jupiter, which proved that Earth was not the centre of all heavenly motions. He was doubtless eager to examine Venus as well, but did not get a chance until after he had

The apparent size of Venus from superior conjunction, when it appears fully illuminated on the other side of the Sun, to near inferior conjunction, when it is approaching its minimum separation from Earth. The variation in apparent diameter is 10″ at left to 60″ at right.

moved to Florence – in September 1610, when Venus was returning to the evening sky after a long absence.[2] Unfortunately, Venus was still very small and far away, and in Galileo's small and optically imperfect telescope little more than a blazing blur of light. Not until mid-November, after Venus reached dichotomy (when its phase would have been half) did the 'horns' of the crescent phase begin to appear. After a few more weeks, Galileo was sure enough to send to Kepler in Prague an anagram (a cipher used to protect priority in discovery), which read: '*Haec immature, a me, iam, frustra, leguntur-o.y.*' (the two letters o.y. were left over). This means, 'I am now bringing

Venus, then an Evening Star, passing through its gamut of phases between January and July 2007. It was captured in this series of images by Sean Walker, using a 31.75-cm Newtonian reflector and a Luminera video camera with a Baader uv filter.

these immature things together in vain.' Once unscrambled, however, the letters become, 'Cynthiae figuras aemulatur mater amorum,' 'The mother of loves imitates the phases of Cynthia [the Moon].'

Venus could not move as Ptolemy had assumed, and Copernicus was vindicated. It was one of Galileo's greatest triumphs, though as is well known, his championing of the Copernican theory got him into serious trouble with the Catholic Church. Sadly, in his late sixties he was summoned to Rome, put on trial and condemned by the Roman Inquisition and forbidden to teach the heliocentric theory. Placed under house arrest at his villa in Arcetri, he remained there until his death in 1642 – one of the great martyrs of science.

Is the Crescent Visible to the Naked Eye?

Detection of the crescent phase is easily within reach of almost any small telescope, and even larger binoculars provided they are stably mounted on a tripod. Controversially, it has been claimed that the crescent can even be made out with the naked eye by exceptionally keen-sighted individuals, which might, if true, account for the way Astarte/Venus was often depicted with crescent horns in antiquity. Sir Henry Layard, who discovered the 'Venus tablet' at Nineveh, also discovered a statue of Astarte bearing a staff tipped with a crescent; there is additionally a figure of Astarte found at Hillah, near Babylon, and now in the Louvre, with a horned headdress. The Dutch historian of astronomy Antonie Pannekoek mentions a text which some Assyriologists have read as 'When Ishtar at her right-hand horn approaches a star, there will be abundance in the country. When Ishtar at her left-hand horn approaches a star, it will be bad for the country.'[3] He adds, 'It does not seem impossible that in the clear atmosphere of [the Mesopotamian] lands the horns of the Venus crescent were perceived.'[4]

The list of observers who have reported making out the crescent with the naked eye is a long one. An American missionary, David

The goddess Ishtar, accoutered with a horned helmet, weapons and a lion, and associated with the Sun. From an Akkadian empire seal, c. 2350–2150 BCE. The horned helmet may have inspired later depictions of Astarte/Venus with crescent horns.

Tappan Stoddard, wrote to Sir John Herschel, the son of the celebrated William Herschel and a noted astronomer in his own right, from Oroomiah in Persia (now Urmia, Iran) in 1852 that 'at twilight Jupiter's satellites and the elongated shape of Saturn could be seen with the naked eye and through a dark glass the half-moon shape of Venus immediately struck the eye'.[5] The elongated shape of Saturn, as well as the sighting of Jupiter's satellites, does not inspire confidence, but there are better-authenticated, if less exotic, reports. The most plausible of these were made with Venus observed either through a dark glass or in twilight, with its dazzling

Venus, Mercury and the 22-hour-old Moon form a right triangle (with Venus at the right angle) on 24 May 2020. Venus was caught as it rapidly moved towards inferior conjunction (on 3 June) and passed within just 12.6′ of arc of the Sun's limb – the margin by which it missed transiting the Sun as it had done one eight-year 'Venus cycle' before, on 6 June 2012. Mercury is somewhat difficult to see; it appears slightly left of centre, in the upper part of the image.

brilliance subdued, and in many cases by younger persons, in whom presumably exceptionally keen eyesight is more likely to be encountered. Thus, according to Patrick Moore,

> The Rev. T. W. Webb relates that the crescent phase was seen without optical aid, before he knew of its existence, by a twelve-year-old boy named Theodore Parker. The phase was undoubtedly visible to E. S. Franks between 1890 and 1900. I can vouch for the accuracy of this, since Franks' father – W. S. Franks, eminent observer and R.A.S. Gold Medalist, who died in 1933 – was well known to me personally.[6]

As against these reports, it should be pointed out that Venus's angular diameter when it shows as a crescent is in the 50–55″ range, which is just about the limit of daylight resolution of the normal human eye (assuming a pupil diameter of 3 mm).[7] Thinking by analogy to the pixels on a display screen, one would presumably have to achieve three-pixel resolution to make out the crescent, which would mean a resolution of more like 15 or 20″. If there are any humans at all capable of attaining this, they must be very rare, and for practical purposes non-existent. The authors can certainly say that we have never seen the crescent with the naked eye, in spite of many attempts. Probably the best that anyone can do is to make out that Venus is elongated at times, as was claimed by Mary A. Blagg, remembered for her work sorting out the nomenclature of the Moon, and the American planetary observer Stephen James O'Meara.[8] As for the ancient association of Astarte/Venus with the crescent, this was likely, as the Lick Observatory's W. W. Campbell concluded, 'a pure lucky guess, probably made under the influence of a crescent moon'.

Planets against the Sun

In 1609, the same year that Galileo first turned a telescope to the sky, Kepler published *Astronomia nova* (New Astronomy). In this book, one of the most important in the history of astronomy, Kepler published a result that he had painstakingly worked out from the precise naked-eye observations of his late employer Tycho Brahe: the shape of the planets' orbits. Instead of the epicycles of both Ptolemy and Copernicus, he found their actual paths are ellipses, in which the Sun is at one focus. In reaching this profound result, Kepler had focused mainly on Tycho's Mars observations, which was fortunate as Mars moves in one of the most highly elliptical orbit of any of the naked-eye planets, after only Mercury. Venus, on the other hand, has an almost perfectly circular orbit; had Tycho chosen to concentrate on Venus, Kepler might never have realized his result.

In the Ptolemaic scheme, Mercury and Venus had been placed between Earth and the Sun. This was, however, merely a convention. It would have been equally reasonable to have placed their orbits beyond the Sun. In the Copernican system, the ambiguity of their position was removed – Mercury and Venus were clearly interior to the orbit of Earth, and Galileo's telescopic observations of the phase merely confirmed this, as noted above. This implied, moreover, that these two planets could sometimes pass directly in front of the Sun, so as to appear as a moving black spot crossing over it. These passages are known as transits, and can only occur at inferior conjunction – that is, when Mercury and Venus lie between Earth and the Sun. Because Mercury's orbit is tilted by some 7° to the plane of Earth's orbit and Venus's by 3.5°, the planets usually miss the Sun, passing either above or below it. Only when an inferior conjunction takes place near one of the nodes – the points where the plane of a planet's orbit intersects that of Earth, known as the ecliptic – can a transit occur. There are two nodes, called the descending node and the ascending node, depending on whether the planet is above (that is, north of) the ecliptic moving downwards or below the ecliptic moving upwards. Transits of Mercury can only occur in early May and early November, and on average, thirteen of them occur each century.[9] For Venus, the corresponding dates occur in early June or early December, but they occur much more infrequently, in pairs separated by more than a century. The last were in 2004 and 2012; the next will not occur until 2117 and 2125 (see Appendix III).

Prior to Kepler, astronomers had relied on the *Prutenic Tables* of planetary motions, calculated by Erasmus Reinhold, a disciple of Copernicus, and the *Alfonsine Tables* based on Ptolemy for their planetary predictions. Kepler, using his elliptic orbits, would produce tables thirty times more accurate that anything that had gone before. As he embarked on the huge project, which would consume him for two decades, he soon realized that Mercury would arrive at inferior conjunction close to one of the nodes on 29 May 1607, and that there was a good chance that a transit would occur.

Setting himself up in an attic in the house in Prague where he was living at the time, and admitting the Sun's light through a crack in the shingles onto a sheet of paper to create a pinhole camera, he began to keep vigil the day before the predicted date of the transit. Thin, scraggly clouds drifted overhead that day, but through intermittent breaks he obtained glimpses of a dark spot on the Sun, which he believed to be Mercury in transit, as he announced in a report on his observations published two years later. However, at the time Kepler did not know of the existence of sunspots, whose discovery waited on the invention of the telescope – they were first recorded in 1611, not only by Galileo, but by several other astronomers at about the same time. At last Kepler realized that he had been mistaken; his calculation had been off, and Mercury did not come to inferior conjunction until 1 June 1607 (rather than 29 May), and he sighed, 'O lucky me, the first in the century to see a sunspot!'[10]

Undaunted, Kepler continued to revise his calculations, and finally published the *Rudolphine Tables*, named after his enthusiastic but often insolvent patron, the Holy Roman Emperor Rudolph II, in 1627, fifteen years after Rudolph's death. While engaged in this Herculean effort, Kepler discovered that a transit of Mercury would occur at midday in Europe on 7 November 1631, followed, coincidentally, within a month by a transit of Venus on 6–7 December. Though Kepler himself died in November 1630, several observers were aware of his predictions and succeeded in seeing Mercury's transit. The most celebrated was Pierre Gassendi, canon of the parish church at Digne, in France, who observed it from Paris. His enthusiasm whetted by this accomplishment, Gassendi made careful plans for the far more spectacular (and rare) Venus transit. At 1 minute of arc, Venus would present a disc against the Sun, six times larger than that of Mercury and visible even with the naked eye. Kepler had predicted that Venus's transit would occur during the European night, but since there were still significant uncertainties in the calculations, Gassendi decided to make the effort to look for it anyway. He put wide margins to his watch, beginning to observe on 4 December; unfortunately, storms raged that day and the next. Only on 6 December – the date for which Kepler had actually predicted the transit – did breaks in the clouds allow Gassendi glimpses of the Sun. He kept up his vigil until three o'clock in the afternoon, but no spot appeared within the circle of his projection screen. On 7 December, he continued his watch through the whole forenoon. Again, there was nothing. In contrast to the transit of Mercury, when there had been several observers, as far as we know, Gassendi was the only one to even try with Venus. He had hoped to add to the glory of his Mercury observation. Fate overruled.

We now know that the transit indeed took place, just as Kepler had foretold. Had anyone looked, it would have been visible from the still-uncharted lands of New Zealand and Australia, from

Asia, Africa and parts of Europe. During a transit of Venus, there are theoretically four so-called contacts: Contact I, when the circumference of the planet's disc moving inwards first comes into contact at a single point with the circumference of the Sun; Contact II, when the planet's disc moving inwards is entirely inside the circumference of the Sun; Contact III, when the planet's disc moving outwards is entirely inside the circumference of the Sun; and Contact IV, when the circumference of the planet's disc last touches the circumference of the Sun at a single point before exiting from it. In central Europe, the transit was already almost over by the time the Sun rose; only Contact IV would have been visible along a swath including Danzig (now Gdansk), Olmütz (now Olomouc), Ingolstadt, Innsbruck, Bologna, Florence, Rome and Naples. Many potential observers may have been distracted by the turmoil of European war. In Paris, by the time the Sun came up, the transit was already over.[11]

A British Triumph

The next transit, eight years later, nearly met the same fate, as Kepler's predictions had not carried forward this far. Fortunately, by then Jeremiah Horrocks, only twenty years old and among Kepler's earliest followers in England, was in the midst of launching a precocious and brilliant, if tragically shortened, career in astronomy.

Horrocks was born in Toxteth, Lancashire, then a small village 5 km from Liverpool and now swallowed up in its suburbs. He received a strict Puritan education, which included a smattering of Latin and Greek. As is often the case, his enthusiasm for astronomy began early, and he later recalled:

When, as a boy, I had first turned my attention to Astronomy, I had many incentives to excite still further my passionate desire. The pleasure of observing used to carry me away in spite of myself, and its sublimity to inspire lofty thoughts. I felt great delight in meditating upon the fame of those pre-eminent masters, Tycho and Kepler, and sought at least to emulate them in my aspirations.[12]

By the age of thirteen, Horrocks had made enough of an impression to be sent to Emmanuel College, Cambridge, where he spent three years and left without taking a degree. The curriculum at the time was almost entirely in Classics; there was no formal instruction in mathematics or physical science at all, and it is likely that he would have received equally good instruction had he simply remained in Lancashire. We do not know why he left, though it seems probable that he could not afford to pay the college the sum required for the degree. Once re-established in Toxteth, he began making astronomical observations, and became acquainted with William Crabtree, a linen draper and amateur astronomer based at Broughton, Manchester, who was just a few years older.

Horrocks's first observations were made with a naked-eye instrument called a radius, which consisted of a bar of wood about a metre long (the radius) at right angles to which was placed a cross-bar with three indices or 'sights', a fixed one in the middle and two others movable, at each end. With this instrument the apparent angular distances between stars could be read off by means of a sight vane. Over the course of about a year, he made scores of observations of the Moon and planets, but on 19 March 1637, a Sunday – the important events in Horrocks's life always seemed to take place on Sunday – he first recorded his use of a telescope, to observe an occultation of the Pleiades by the Moon.

Two months later, Horrocks obtained a copy of the *Rudolphine Tables*. By means of his own and Crabtree's observations he confirmed that Kepler's theories were accurate, but also saw that the predictions from them could be improved by adjustment of some of Kepler's numerical constants. Thus began the great work which

Horrocks window in the north wall of St Michael's and All Angels Church, Much Hoole, commemorating his 4 December 1639 observation from near this location. This is, of course, a modern conceptualization, and no contemporary portrait of Horrocks exists. The window was installed in 1872.

was to occupy the remainder of his short career. During what was to be for him the great year of 1639, he moved from Toxteth to the small village of Much Hoole, a rather desolate site 16 km southwest of Preston, of which his nineteenth-century biographer the Rev. Robert Brickel says, 'though doubtless an open situation for an astronomer, it could not have been a very agreeable residence'.[13] Apparently, Horrocks had some kind of position at the Anglican church of St Michael's, though he was not a curate, as pious Victorians liked to imagine; perhaps he was a priest's assistant or a schoolmaster. In any event, in the autumn of that year, he was busy calculating tables of the Sun, Venus and Mars when

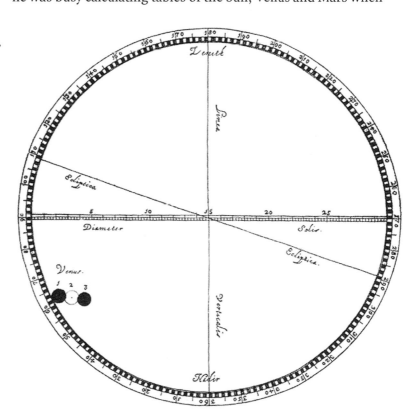

Horrocks's diagram showing his observations of Venus having just entered the Sun's disc before sunset, at the transit of 4 December 1639. From Johannis Hevelii, *Mercurius in Sole Visus Gedani* (1662).

he made a remarkable discovery – something Kepler had missed. In a letter written on 26 October 1639 (Old Style; 5 November New Style) he informed Crabtree of 'a remarkable conjunction of the sun and Venus on 24 November next [Old Style; 4 December 1639 New Style], on which day the planet will transit the solar disc. Such an event has not, indeed, happened for many years past, nor will it occur again this century. I, therefore, earnestly beg of you to apply yourself diligently to your telescope, and make whatever observations you can.'[14]

As the event was only a month away, there was little time to communicate his calculations to others. Apart from Crabtree, the only other person who seems to have been aware of what was about to take place was Horrocks's brother, Jonas, in Liverpool. The eagerly awaited day was a Sunday, and Horrocks records in the book he later wrote about the transit, *Venus in Sole Visa*, that he watched carefully from sunrise to nine o'clock, and from a little before ten until noon. Then, at one in the afternoon, he was called away 'by business of the highest importance which . . . I could not

Naked-eye view of Venus in transit, 6 June 2012. View at sunrise from near Darnózseli, Hungary.

with propriety neglect'.[15] What business could possibly take priority for an astronomer over an event that would not occur again for 122 years? Presumably, it was attendance at divine services, which during this religion-obsessed age were regarded as mandatory, without exception. But clouds were also a worry. Fortunately, at 3.15 p.m., just 35 minutes before sunset on a short winter day, the clouds dispersed just as he resumed his place at his telescope, and as he tells us: 'I . . . beheld a most agreeable spectacle, the object of my sanguine wishes, a spot of unusual magnitude and of a perfectly circular shape, which had already fully entered upon the Sun's disc.'[16] Crabtree at Broughton was also successful, but Jonas was clouded out. Only two men, Horrocks and Crabtree, succeeded in witnessing the only transit of Venus to be observed during the whole seventeenth century.

Though he accomplished other important work, such as improving the theory of the Moon's motions and discovering the 'Great Inequality' in the motions of Jupiter and Saturn – the fact that Jupiter was moving more rapidly and Saturn more slowly than Kepler's values for the mean motions implied – Horrocks will always be best remembered for the prediction and singular observation of 4 December 1639. Sadly, he did not have long to live. On 3 January 1641, another Sunday, while organizing a visit to see Crabtree, he died very suddenly, of unknown causes. He was barely 22. One can only imagine what he might have accomplished had he lived a longer life. But it was not to be. His biographer, Sidney Bertram Gaythorpe, applies to Horrocks's early death lines from John Milton's poem 'Lycidas' (written in 1637 on the death of one of Milton's Cambridge classmates):

But the fair guerdon when we hope to find,
And think to burst out into sudden blaze,
Comes blind Fury with abhorréd shears
And slits the thin-spun life.[17]

THREE

Adventures in the Seventeenth and Eighteenth Centuries

The revelation of the phase is exciting for the fledgling astronomer, but the sight of it grows familiar, and one soon wishes for more. There is every reason for high expectation. As a 'star' its beauty is beyond compare: with a magnitude of as much as −4.7, it is far brighter than any other planet or star; it is often readily visible in broad daylight and inspired perhaps the vast majority of UFO reports. According to the French astronomer François Arago, it even managed to upstage Napoleon on one occasion in 1797:

> [Alexis] Bouvard has related to me that General Bonaparte, upon repairing to [the Palais de] Luxembourg, when the Directory was about to give him a fête, was very much surprised at seeing the multitude which was collected in the Rue de Tournon pay more attention to the region of the heavens situate above the palace than to his person or to the brilliant staff which accompanied him. He inquired of the cause, and learned that these curious persons were observing, with astonishment, although it was midday, a star, which they supposed to be that of the Conqueror of Italy; an allusion to which the illustrious general did not seem indifferent when he himself with his piercing eyes remarked the radiant body. The star in question was no other than Venus.[1]

Venus together with Halley's Comet, photographed by Earl C. Slipher in May 1910. During the exposure, the camera was guided on the comet and Venus; the lights of Flagstaff, seen on the right side of the plate, trailed during the exposure.

The 'star of Napoleon'.

The reality is, in fact, disappointing. The nineteenth-century astronomer Sir John Herschel (1792–1871) remarked that in the telescope Venus shows a brilliance that

dazzles the sight and exaggerates every imperfection of the telescope . . . we see clearly that its surface is not mottled over with permanent spots like the Moon; we notice in it neither mountains nor shadows, but a uniform brightness, in which sometimes we may indeed fancy, or perhaps more than fancy, brighter or more obscure portions, but can seldom or never rest fully satisfied of the fact.[2]

Another great nineteenth-century observer, William Frederick Denning (1848–1931), wrote in a similar vein:

When the telescope is directed to Venus it must be admitted
that the result hardly justifies the anticipation. Observers are
led to believe, from the beauty of her aspect as viewed with the
unaided eye, that instrumental power will greatly enhance the
picture and reveal more striking appearances than are displayed
on less conspicuous planets. But the hope is illusive. The lustre
of Venus is so strong at night that her disc is rarely defined with
satisfactory clearness; there is generally a large amount of glare
surrounding it, and our instruments undergo a severe ordeal
when their capacities are tested upon this planet.[3]

In a non-achromatic telescope, false-colour effects abound, and
instruments that are poorly aligned and involve a series of reflecting
surfaces will produce an array of secondary images ('ghosts'). Such
effects led to various reports in the eighteenth century that Venus had
a moon; though some of the observers were naive, rather surprisingly
even the great Giovanni Domenico Cassini (1625–1712) was thus
deceived.[4] Apart from employing excellent optics, an observer will
find more satisfactory views at twilight, when the planet's brilliance
is reduced and the eye less dazzled. Serious work, however, is best
attempted during daylight periods. Even at those times much is to
be said for the employment of a neutral-density or variable-density
polarizing filter to reduce the glare. The nineteenth-century English
instrument-maker George H. With employed a Newtonian reflector
with an unsilvered mirror for all his Venus work.

The at first uninspiring view of the planet was described by
the pioneering observer Christiaan Huygens (1629–1695), best
remembered for his discovery of the ring round Saturn. Near the
end of his career he summed up some of the conjectures that were
suggested by his observations. He had, he wrote,

often wonder'd that when I have viewed Venus at her nearest to
the Earth, when she resembled an Half-moon, just beginning

to have something like Horns . . . she always appeared to me all over equally lucid, that I can't say I observ'd so much as one spot in her, tho in Jupiter and Mars . . . they are very plainly perceived. For if Venus had any such thing as Sea and Land, the former must, necessarily, show much more obscure than the other, as any one may satisfy himself, that from a very high Mountain will but look down upon our Earth. I thought that perhaps the too brisk Light of Venus might be the occasion of this equal appearance; but when I used an Eye-glass that was smok'd for the purpose, it was still the same thing. What, then, must Venus have no Sea, or do the Waters there reflect the Light more than ours do, or their Land less? Or rather (which is most probable in my opinion) is not all that light we see reflected from an Atmosphere surrounding Venus, which being thicker and more solid than that in Mars or Jupiter hinders our seeing anything of the Globe it self, and is at the same time capable of sending back the Rays that it receives from the Sun.[5]

In 1666–7 Cassini, observing at Bologna, Italy, before being called to the Paris Observatory founded by Louis XIV, claimed to make out a bright spot on several occasions, of which he said that it completed its motion, 'whether of Revolution or Libration, so as in near 23 days it returns about the same hour to the same situation in the planet; which yet happens not without some irregularity.' The meaning of this is not very clear.[6] However, in 1740, Cassini's son, Jacques, later reworked the observations to arrive at a value of 24 h 20 m. The analogy to Earth seems henceforth to have proved irresistible, and for the next two centuries, with only one exception, a period around 24 hours prevailed.

The exception was Francesco Bianchini, who was born in 1662 into a noble family at Verona. Pursuing a career in the Church, he became librarian to Cardinal Pietro Ottoboni, and when the latter became Pope Alexander VIII, rose to the rank of papal chamberlain.

He corresponded with Cassini, and while on a diplomatic mission to Paris in 1712, gave the great astronomer, who was then dying, the last rites of his Church. The following year he was elected a Fellow of the Royal Society of London, his name being put forward by Sir Isaac Newton. In addition to astronomy, he was a keen student of Roman antiquities, and in August 1725, just before he began the studies of Venus for which he is now remembered, he suffered a serious mishap while exploring the ruins of the Palace of the Caesars on the Palatine Hill, which almost interrupted the observations before they began:

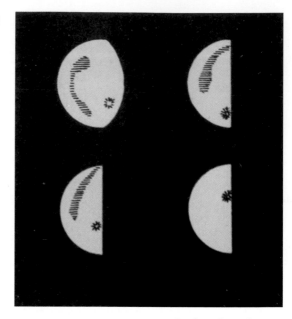

Drawings of Venus by G. D. Cassini, made in Bologna, Italy, 1666–7.

> While I was carelessly running about to take the measurements of the surviving rooms in the east wing of Augustus' home . . . I fell into a broad, deep hole in the pavement which I had not noticed . . . I broke my right thigh, and by God's mercy was only saved from the imminent death which threatened me by pressing with all my power with both hands and my left foot against the sides of the hole, to sustain the weight of my body and avoid falling headlong into the pit 40 palms [8 m] below.[7]

Fortunately, by the beginning of the year 1726 he had made a full recovery, and in February and March commenced his studies of Venus as an Evening Star, using 'aerial' telescopes of 60-mm aperture and 19.5- and 20-metre focal length, with a magnifying power of 112.[8] He set up his 'machines' in various places from which a suitable view of the western sky was available. One such place was a garden of the Palazzo Barberini on the Quirinal Hill, another was

An 'aerial' telescope constructed by the Roman instrument-maker Giuseppe Campani, for the use of G. D. Cassini at the Paris Observatory.

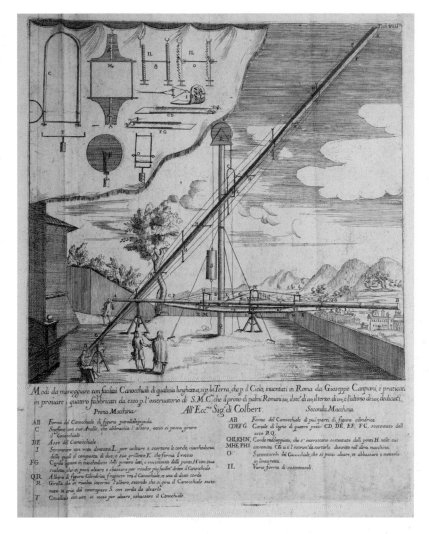

a somewhat smaller site on the Esquiline Hill. However, most of the observations were made at Albano, 25 km outside Rome. Setting to work in twilight about a half hour after sunset, and working for about an hour until he ran out of observing time before Venus set, Bianchini succeeded in making out a series of dusky spots along the terminator of the planet, which resembled the 'seas' in the Moon as seen with the naked eye, though less distinct.

49

Planetæ *Veneris* Phases conspectæ, et maculæ ad instar *Marium* Lunarium detectæ, necnon vertigo circa *Axem* proprium spatio dierum 24 Romæ et Albani annò 1726 per Telescopia I. Campani Palmorum Rom. 88 et 94.

die 9 Februarii

die 24 Febr.

TAB. I

die 16 Febr

die 18 Febr.

Omnia exhibentur situ inverso, ut in telescopio apparebant:
Detexit, observavit, edidit.
Franciscus Blanchinus Veronensis S.S. D.N. Papæ Prælatus Domesticus

Bianchini adjusting the eyepiece end of a 'machine' (aerial telescope) while observing Venus.

Venus observations by Francesco Bianchini, made with telescopes of 88 Roman palms (19.5 m) and 94 Roman palms (20 m) focal length, at Rome and Albano during the evening elongation of February 1726.

He observed again at the morning elongation of May to June 1726, at the evening elongation of July to September 1727, and finally at the morning elongation of January 1728. At last he was ready to publish his results in a book: *Hesperi et phosphori nova phaenomena, sive observationes circa planetam Veneris* (New Phenomena of Hesperos and

Phosphoros, or Rather Observations Concerning the Planet Venus), which he dedicated to the king of Portugal, Dom John V.

What are we to make of the dark spots shown in Bianchini's drawings? The contrast is no doubt exaggerated, but they cannot be dismissed out of hand. There is certainly no doubt that Bianchini was a skilful observer, who in addition to his work on Venus made the first drawing of the Alpine Valley on the Moon and discovered three comets. The Italian skies were excellent, and as is now known from optical analysis of lenses of the period, his telescope permitted

A globe of Venus, based on the map published by Bianchini in *Hesperi et phosphori nova phaenomena* ... (1728). Bianchini gave the study of the markings he believed he had discovered on Venus the rather ugly name of celidography, from the Greek words for 'mark' and 'to write'. The features shown here, Mare Constantini and Mare Emmanuelis, honour the Portuguese rulers King Emmanuel, 'The Fortunate' – who, wanting to be king of Spain as well as of Portugal, expelled the Jews, the backbone of Portuguese commerce, to please the Spaniards and Prince Constantine of Braganza, Viceroy of Goa.

transmission of the short-wavelength part of the spectrum down to about 325 nanometers (nm), where the venusian features are known to be most prominent.[9] There is every reason to believe what he saw was genuine. On the other hand, his inferences were not so happy. He believed the spots to be definite and permanent enough to yield a rotation period of 24 days, 8 hours, and imagined them to represent various topographical features, such as oceans and seas. He even drew up a map, constructed out of gores so it could be assembled into a globe, and gave the features the names of Catholic explorers such as Columbus and Vespucci, the astronomer Galileo (who despite his differences with the Church died a good Catholic), and princes and kings such as Henry the Navigator and Bianchini's own patron Dom John v of Portugal (known as o Magânimo, the Magnanimous). A horseshoe-shaped marking near the presumed north pole was named the Marco Polo Sea; another 'sea', near the south pole, was named for Ferdinand Magellan and so forth. As late as the nineteenth century, the English clergyman and eminent amateur astronomer T. W. Webb encouraged his readers to transfer the markings on Bianchini's map to a small globe.[10] Whatever one may say of the reality of the spots, the way they are disposed on his globe leads to an illusory result, as they were produced from a completely arbitrary combination of the drawings.

Venus's Transit in 1761

At the same moment that Bianchini was making his careful telescopic observations of Venus, the planet was beginning to preoccupy the thoughts of the young astronomer Joseph-Nicolas Delisle. Briefly ensconced in an observatory in the cupola of the Palais de Luxembourg (until evicted by the Duchesse de Berry when she took up residence), Delisle was named to the chair of mathematics at the Collège Royal, a position that Pierre Gassendi had once held. A compulsive letter writer, he was in regular contact with Edmond Halley, whom he had met during a visit to England in 1724. He became fascinated to the point of obsession with the latter's method of using the transits of Venus to measure the astronomical unit, the distance from Earth to the Sun.

Halley – who as a 21-year-old student at Oxford had taken leave of absence to map the southern stars from the island of St Helena, where Napoleon was later exiled – had observed a transit of Mercury on 7 November 1677.[11] As he watched it, Halley realized at once that the geometry of the transit was such that if two observers were widely separated in latitude, they would see the planet move along different chords as it traversed the Sun. The lengths of the chords could therefore be determined (or, alternatively, the time taken by the planet to cross the Sun). This would give the shift in the Sun's observed position due to the difference in the observers' positions. So long as the distance between the two observers was known, the distance to the planet could be worked out, and by application of Kepler's third law, the distance to the Sun itself. In practice, the solar distance is often expressed in terms of the so-called solar parallax, the angle subtended by Earth's radius as viewed from the Sun.

Mercury, however, is rather small and too far away for the observations to be taken with the necessary accuracy, and Halley recognized that Venus offered a much better subject. Although he published a paper titled 'On the Inferior Conjunctions of the Inferior Planets with the Sun' in 1691, explaining how 'by this observation alone, the distance of the sun, from the earth, might be determined with the greatest certainty', only in 1716, at the age of sixty, did he publish a paper in the *Philosophical Transactions of the Royal Society of London* going into the practical details and confidently claiming that observations of the contact points of Venus's disc with the limb of the Sun could be used to determine the distance from Earth to the Sun to an accuracy of 1 part in 500. As with his prediction of the return of the comet that would come to bear his name, Halley knew that he himself would never live to see the fulfilment of his anticipations; he died in 1742. The next transits of Venus would not occur until 6 June 1761 (nearly 122 years after that observed by Horrocks and Crabtree) and 3 June 1769. Nevertheless, the timing was impeccable. It was the age of the vast European empires. The British, from 1685, had gained far-flung colonial possessions (as had Spain since 1570, Holland since 1612 and Portugal since 1674). In addition to their rivalries in commerce, these empires competed for the prestige derived from exploration and scientific discovery, rather as the USA and USSR were to compete during the Space Age. In order to apply Halley's method successfully, a worldwide network of observers would be needed, and this meant, despite the rivalries, coordinating the observations on an international scale.

This is where Delisle came in. Among his numerous correspondents was Peter the Great, Emperor of all Russia. Ambitious to modernize Russia along Western lines, in 1725 Peter invited Delisle to St Petersburg to found a school of astronomy and an observatory. Delisle accepted, but Peter died before the plan could be set in motion. The offer was renewed by Peter's second wife, the Lithuanian peasant-turned-empress consort and now empress regnant of Russia Catherine I (not to be confused with Catherine II, called 'the Great'). Delisle may not have expected to stay long; the Collège Royal granted a leave of absence of four years during which he could still reclaim his position. As it turned out, he would remain in Russia for over twenty years. On his return to Paris in 1747, Louis XV, wanting to ensure that he stayed put this time, appointed him 'Astronome de la Marine' and granted him a

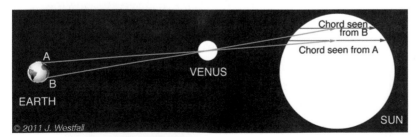

Halley's method for determining solar parallax: chords across the Sun from two stations, A and B, widely separated by latitudes within the zone from which the transit is visible in its entirety, traverse chords of different lengths (follow paths of different durations) when crossing the Sun's disc. The method was later modified by Joseph-Nicolas Delisle so that observers would only need to observe the beginning or end of a transit.

lifetime annuity. With this support, the astronomer established his base of operations in the Hotel de Clûny on the West Bank near the Sorbonne.

In contrast to Bianchini, Delisle had little interest in Venus as a world. His interest was entirely in Venus's transits, and in the application of these rare events to determining the basic unit that would define the scale of the solar system. Possessed of demonic energy, he set himself up as organizer-in-chief of all French transit expeditions, and in addition he identified a serious, and potentially fatal, flaw in Halley's original method. Though it only required timing the duration of the transit (or the length of the chord of the planet's track), it could only be applied from those locations where the transit was visible from beginning to end. In addition, to give the best results, the wider the separation in latitude the better, and though obviously most observers would remain in place in Europe, at least some expeditions would have to be sent to very remote locations. There was the additional problem with Halley's method in that the interference of clouds at either ingress or egress would render the observations useless. Realizing these difficulties, Delisle devised an alternate method in which observers at a pair of stations would both time the same phase of the transit, say the beginning or the end. This increased the probability that the weather would cooperate, though inevitably there was a trade-off, as Delisle's method required the observers to work out the exact longitude of their station.

Delisle's extensive preparations included production of a series of beautiful mappemondes (maps of the world showing the two hemispheres as separate circles) showing the optimal locations from which to make observations. On the basis of his recommendations plans were set in motion to send the abbé Jean-Baptiste Chappe d'Auteroche to Siberia, Guillaume-Joseph-Hyacinthe-Jean-Baptiste Le Gentil de la Galaisière to Pondicherry in India and Alexandre-Gui Pingré to the island of Rodrigues east of Madagascar. Though slower in getting organized, the British were also active. A committee of the Royal Society of London under the leadership of the future Astronomer Royal at Greenwich Nevil Maskelyne drew up plans to send expeditions to St Helena, to be led by Maskelyne himself, and to Sumatra, where the observers were to be Charles Mason and Jeremiah Dixon. Unfortunately, before the transit took place, the Seven Years War between Great Britain and France broke out, and the travel plans of some of the astronomers were upended. The harrowing and often protracted adventures of the participants have become legendary.[12] Some of the observers did obtain useful results. Chappe, for instance, who enjoyed fine weather in Siberia, made good timings, and also noted the unexpected appearance of a shining arc, or 'little atmosphere' (aureole), round the planet's disc. It appeared just after egress and remnants remained visible until three-quarters of the diameter of Venus's disc lay outside the Sun. Pingré, however, was not so fortunate; it was raining at the beginning of the transit at his station on Rodrigues Island, though the weather improved and he was able to observe the later stages and make timings that allowed the application of Delisle's (though not Halley's) method.

The most unfortunate of all was Le Gentil. Surpassing the others in diligence, and travelling at his own expense, he had set out early, and arrived at Isle de France (now Mauritius) a full year in advance of the transit. Almost at once, however, there were forebodings of troubles to come. Karaikal, a French settlement on the Coromandel coast south of Pondicherry, had been captured by the British, and Pondicherry itself was blockaded and under siege. A French fleet, forming at Isle de France and intended to relieve Pondicherry (and so provide Le Gentil with a means of reaching his destination) was wrecked by an unexpected hurricane in January. Le Gentil was on the point of abandoning his original plan of observing from Pondicherry and instead considered shipping out for Batavia in the Dutch East Indies. While trying to decide, he

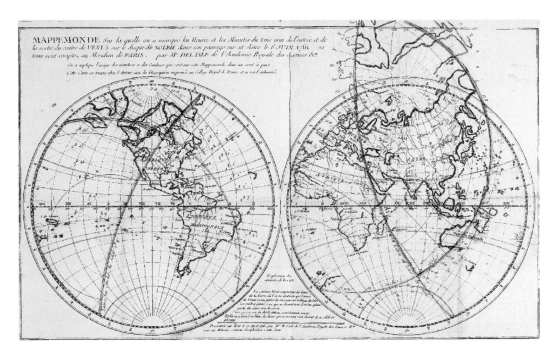

Delisle's mappemonde for the 1761 transit of Venus.

came down with dysentery. It was March by the time he was fit to travel and another French ship, *Le Sylphide*, was available, and ready to set sail in another attempt to raise the siege of Pondicherry. Le Gentil thought success promising enough to stick to the original plan. Unfortunately, his bad luck continued; the ship was blown off course, and as he wrote afterward,

> we wandered . . . for five weeks in the seas of Africa, along the coast of Ajan [Aden], in the Arabian seas. We crossed the archipelago of Socotra, at the entrance of the gulf of Arabia . . . We appeared . . . on the coast of Malabar, the 24th of May; we learnt from the ships of this country that this place was in the possession of the English, and that Pondicherry no longer existed for us . . . I would not yet have despaired if we had followed our first objective to go to the coast of Coromandel; but they made, to my great regret, the decision to return to the Isle de France.[13]

57

The rerouted vessel failed to arrive in time. Instead, when the transit occurred, Le Gentil found himself somewhere in the middle of the Indian Ocean. Though he observed Venus's passage across the Sun from the bridge of the ship, since his precise longitude was unknown, his data was completely useless for the purpose for which he had travelled so far.

How the Black Drop Spoiled the Observations

Halley had assumed the contacts – the points where the limbs of Venus and the edge of the Sun barely touch – to be geometrically precise. This implied a 1-in-500-part accuracy in the determination of the astronomical unit. However, the actual observations were confounded by an optical effect that produced an indistinctness, blurring or strip of darkness (known as the 'black drop') on Venus's image when adjacent to the limb of the Sun. Not only was the effect ubiquitous among observers, it was also highly variable, and so could not be readily controlled for. Indeed, timings even by observers standing side by side might differ by a minute or more.

The cause of the black drop is now well understood: it is owing to blurring of images caused by tremulous air and diffraction of light in the telescope, with a smaller but non-negligible effect owing to the falling off of light near the Sun's limb. In the past, many writers wrongly described it as an effect of Venus's very considerable atmosphere, though this is clearly incorrect, since a black drop effect is also observed at transits of airless Mercury.[14]

A quite different phenomenon is the bright aureole seen round Venus's disc just before it enters onto or emerges off the limb of the Sun, produced by refraction of sunlight by Venus's atmosphere. Its discovery has been widely credited to the famous Russian academician Mikhail Vasilevich Lomonosov.[15] This has been doubted, and Chappe may have been first, though in fact the matter of priority is of no real importance.[16] It is extremely bright,

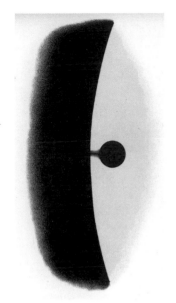

The infamous 'black drop', which ruined the precision of many transit observations. This is a view by the 19th-century observer George Hirst of New South Wales, Australia, at the 9 December 1874 transit of Venus. Hirst was using only a 4-cm refractor, but in 'remarkably clear and steady atmosphere'. Poor seeing will generally enhance the effect.

The aureole, a brilliant but delicate arc of light seen along the limb of Venus before it has entered upon or after it has exited the solar disc, first seen by observers of the 1761 and 1769 transits. Observation by A. J. Park, at Armidale in the Northern Tablelands of New South Wales, using a 12-cm refractor at the 9 December 1874 transit of Venus.

as it represents an actual refracted image of the Sun, but also quite delicate, since the total apparent angular height of Venus's air is only about 0.02 arc second.

As we now know, the aureole is produced in the mesosphere of Venus's atmosphere, the layer above the upper cloud deck; detailed observations at the modern transits of 2004 and 2012 have even been used to work out the vertical structure of Venus's atmosphere.[17]

More Travellers' Tales: 1769

Though the 1761 results were disappointing, largely because of the unexpected interposition of the black drop, astronomers could regard it as something of a dry run, since no one living had ever seen such an event. They now had eight years to plan ahead, and at the next transit they could expect to be better prepared. In contrast to 1761, when the French had been on their game far sooner than the British, at the 1769 transit the roles were reversed; the British now took a decisive lead over their continental rivals.

William Wales and James Dymond headed to Prince of Wales Fort (now Churchill) on Hudson Bay in Canada. After surviving a bitter Canadian winter, they were rewarded with a clear sky on transit day, and obtained excellent measurements of the contacts. Meanwhile, in what was by far the most ambitious enterprise of the time, James Cook sailed the *Endeavour* halfway round the world to observe the transit from the just-discovered tropical paradise of Tahiti in the South Seas. In searing heat on transit day, Cook himself assisted the expedition's astronomer Charles Green, and made successful observations from Point Venus, named in honour of the occasion. After the transit he took up his further charge to locate new southern lands, observe the potential of their natural resources, the 'Genius, Temper, Disposition and Number of the Natives', and claim them for King George III (who would also observe the transit, from his private observatory at Kew). Thus began the first of Cook's three celebrated circumnavigations of the globe.

But the French also remained active. Le Gentil, rather than return home to France, decided to wait out the eight-year intermission south of the equator until the next transit; he devoted himself to the study of the botany, zoology, geology and anthropology of Mauritius and Madagascar. However, his ill luck with Venus continued. He planned to observe the transit from Manila, but was ordered to proceed to Pondicherry – which in the eight years since the last transit had gone from British back to French control. He followed orders, and suffered the astronomer's most dreaded fate: he was clouded out. After ten years of voluntary exile from France, there was nothing left for him to do but return home. (It is pleasant to record that his later years were happy ones; he married a wealthy heiress named Marie Potier, had a daughter, Marie Adélaïde, upon whom he doted, and lived in comfort at the Paris Observatory until his death in 1792, a year before the Reign of Terror of the French Revolution began.)

At least he returned safely. Not so Chappe. The successful veteran of the Siberian expedition of 1761 led another, including five Frenchmen and two Spanish officers, to a Spanish mission at San José del Cabo at the tip of Baja California for the 1769 transit. The second Chappe expedition typifies the lengthy journeys and unforeseen delays endured by eighteenth-century overseas expeditions: although they left Paris eight and a half months before transit day, they arrived at San José del Cabo with barely enough time to set up their instruments. They also arrived to find that an 'epidemical distemper' (probably typhus) was raging through the mission. It had already killed one-third of the local population, but there wasn't time to take up a different position. Clear skies on transit day meant success for Chappe. He and several colleagues recorded all the contacts, and Chappe again recorded the arc. But because of their late arrival they had not yet determined their longitude. Now, one by one, they began to fall victim to the epidemic. Despite being ill, Chappe, with remarkable determination, managed to time the lunar eclipse of 19 June; other remnants of the party timed several eclipses of the Galilean satellites, and so the longitude was found. Chappe hung on until 1 August, when he died a true martyr to science. Of the original party, only two survived to bring the observations back to France.

So ended the great eighteenth-century transit expeditions, though the results would not be fully analysed for years, indeed decades, to come.

FOUR

A Planet of Illusions – and a Few Facts

Hesper-Venus – were we native to that splendour
     or in Mars,
We should see the Globe we groan in, fairest of
     their evening stars.
Could we dream of wars and carnage, craft and
     madness, lust and spite,
Roaring London, raving Paris, in that point of
     peaceful light?

ALFRED, LORD TENNYSON,
LOCKSLEY HALL, SIXTY YEARS AFTER (1886)

As Venus passed off the solar disc on 3 June 1769, it left a legacy of observations that would preoccupy astronomers for decades. Not only did all the timings need to be collated but the precise locations on Earth's surface where they were taken had to be calculated. Further, each observation had to be weighted based on the level of confidence in the observers, their instruments and the seeing conditions at the time. Preliminary results (independently arrived at) were published by Thomas Hornsby and Jérôme Lalande in 1771, and by Anders Lexell and Alexandre-Gui Pingré in 1772. Their values for the solar parallax ranged from 8.58″ to 8.80″. To convert these values to distances in kilometres (or miles), an accurate value of Earth's equatorial diameter was needed. However, an authoritative result was long delayed; it was not published for another half century, at almost the midpoint between the eighteenth-century transits and the next pair predicted for 9 December 1874 and 6 December 1882.

The backbreaking work of the huge calculation was undertaken by a German astronomer and former artillery officer in the Prussian army, Johann Franz Encke. In 1822 he became director of the Seeburg Observatory at Gotha, Germany, and over the next two years he calculated and recalculated the data from the eighteenth-century transits. His final value for the solar parallax, published in 1824, was 8.5776″, with a stated uncertainty of only 0.4 per cent. Combining this with an accurate determination of Earth's equatorial diameter (at the time the best available was by the Finnish geodesist Henrik Walbeck in 1819), the astronomical unit was found to be 53,340,000 ± 660,000 km, or 95,280,000 ± 410,000 miles, a value adopted by the British *Nautical Almanac* and the *American Ephemeris*, and memorized by a generation of schoolchildren. By the time it was retired, in 1869, preparations for the 1874 and 1882 transits were far advanced, and astronomers were optimistic that with advances in transportation, instrumentation and observational techniques, including the introduction of photography, a much more accurate result would be achieved.

Venus as a World

All the attention to the transits of Venus, in which the planet was used only as a target for measuring the distance to the Sun, led to neglect of the study of Venus as a world. In part, this neglect was due not only to the fact that astronomers' efforts were otherwise directed but to the great difficulties involved in studying Venus's dazzling and nearly featureless orb, from which a highly reflective cloud deck casts 80 per cent of the incoming sunlight back into space. This ubiquitous cloudscape presents at first glance an almost

invariably featureless surface to a visual observer, on which at most
a few nebulous patches of uncertain outline and little contrast can
be distinguished. The casual observer is defeated at once, while
even diligent observers struggle mostly in vain. The verdict of the
renowned late nineteenth-century American astronomer Edward
Emerson Barnard, who in 1897 summed up his experiences with
the 30.5- and 91-centimetre refractors at Lick Observatory, has
been seconded by most visual observers of Venus. 'The planet was
watched through many years,' he wrote, 'but indifferent seeing
always baffled [me] and no satisfaction could be gotten out of it
. . . Vague suggestions of spots were frequently present.' He added,
'We have – considering the promise its great brilliancy holds
out – perhaps one of the most disappointing objects in the whole

Venus as drawn by the
great American planetary
observer E. E. Barnard
on 29 May 1889, with the
30.5-cm refractor at Lick
Observatory. The air was
thick with smoke from
a wildfire burning in the
canyon near Mt Hamilton,
which acted as a kind of
natural neutral-density
filter to improve the view.
Barnard considered this the
best view of the planet he
ever had.

heavens.'[1] Under the circumstances, very little could be found out about the planet. Its diameter and mass were known, and showed it to have similar dimensions to Earth. It had an atmosphere. However, despite numerous conjectures, even the rotation period and the inclination of the axis long remained uncertain.

Until the last decades of the eighteenth century, Bianchini's on the whole rather doubtful observations still held sway. However, the study of Venus the world resumed with the determined efforts of two observers of skill and enthusiasm, William Herschel and Johann Hieronymus Schroeter.[2] In the 1770s Herschel was a musician with a passion for astronomy and telescope-making who, taking advantage of the 'personal union' between the House of Hanover and the throne of Great Britain between the accession of George I in 1714 and that of Victoria in 1837, had emigrated from Hanover to England. Schroeter was secretary of the Royal Chamber of King George III in Hanover; his own interest in music had placed him in touch with Herschel's brothers, and in 1779 this led to the acquisition of a small achromatic refractor with which he observed Venus 'not only daily, but, as far as the weather and her position admitted, almost hourly through the whole day and evening'.[3] It is apparent from this that Schroeter was one of the first, if not the very first, to observe Venus during broad daylight, rather than only during the twilight periods. Herschel's discovery of the *Georgium Sidus* ('Star of George', the planet now known as Uranus) in March 1781 sealed the fates of both men. Supported financially by George III, Herschel was able to retire from music. With the collaboration of his sister Caroline, he was henceforth able to devote all of his time to astronomy and telescope-making as the Court Astronomer (not the same as Maskelyne's position of Astronomer Royal), while Schroeter, in a spirit of emulation, gave up his demanding position in Hanover for that of chief magistrate of the village of Lilienthal, near Bremen, which allowed him to devote much of his time to astronomical work. Both men made

some observations of Venus, but Schroeter was by far the more assiduous. A number of critical observations established the existence of a dense cloudy atmosphere. For instance, he noticed that the light falls away very markedly towards the terminator (the line between the day and night hemispheres of the planet), so that the time that Venus appears half to the eye differs by several days from the theoretical date (this is sometimes called the 'Schroeter effect'), which he correctly interpreted as due to absorption in the planet's atmosphere. He was also the first to notice that near the time of inferior conjunction, when the angular distance between Venus and the Sun is less than about 2°, the cusps of the slender crescent are extended for some distance round the dark side and may even coalesce into a delicate aureole encircling the entire disc. Also due to Venus's atmosphere, it is produced by forward-scattering of sunlight in the upper atmosphere of Venus above the cloud layer, the same process that produces twilight on Earth. These observations put beyond doubt Christiaan Huygens's long-ago supposition that the planet's dazzling appearance is due to the presence of a cloudy atmosphere.[4]

All this was good, solid work. A more doubtful inference proceeded from an observation he made on 28 December 1789, while using a 2.1-metre focal-length reflector of which the optics had been acquired from Herschel. He detected a series of indentations along the terminator and in addition found that the southern horn was blunted, as if it had been pared away, while a tiny speck of sunlight was noted trembling in the darkness beyond. He

Extensions of the cusps of Venus near inferior conjunction in June 1964. The images were obtained by A. Dollfus and E. Maurice at the Pic du Midi Observatory, and from left to right were taken on 17 June, 17 June, 18 June, 21 June and 25 June. Note the rapid changes in orientation of the slender crescent from one day to the next.

The cusp extensions. Drawings by E. E. Barnard using the 30.5-cm refractor at Lick Observatory. The planet was observed during daylight. Notice the thready appearance of the ring.

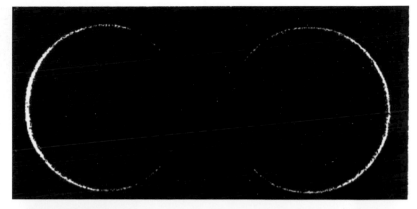

A similar thready appearance to the ring seen by Barnard near inferior conjunction appears in this CCD image by Sebastian Voltmer on 2 June 2020.

inferred that this quivering apparition bore witness to the existence of a towering mountain, 40 km high, at this location, and believed in its reality for the rest of his life – weathering even the harsh criticisms of Herschel himself, who chided, 'As to the mountains in Venus, I may venture to say that no eye, which is not considerably better than mine, or assisted by much better instruments, will ever get a sight of them.'[5] From the fact that the blunted-horn appearance was nearly exactly repeated from night to night, Schroeter determined that the length of the rotation period was just a little less than Earth's, giving a value of 23h 21m 19s in 1793, which was later 'corrected' to 23h 21m 7.977s. These values were long cited in the literature as authoritative.

Thus to Bianchini's seas and continents was added another strand to the increasingly intricate lore of the planet – what the indefatigable Venus observer and historian of astronomy Richard Baum has called the 'Himalayas of Venus'. In describing their fascination on a succession of observers and writers in the decades after Schroeter introduced them, Baum concludes:

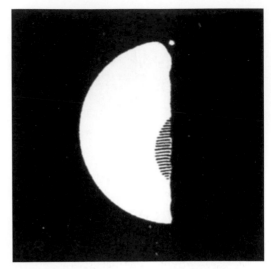

The 'enlightened mountain', a sketch based on Johann Hieronymus Schroeter's observation of 28 December 1789. The original drawing was published in Schroeter's *Selenotopographischen Fragmenten über den Mond*, Band (1791).

> And yet nothing was what it seemed. The cordilleras, the icy peaks, the summit ridges, craters, chasms and gorges, all the geological forms known to humankind that tripped in glittering pageant from the pens of astronomical observers were as incorporeal as the landscapes of cloudland – shadowy places where memory and imagination, not strata of rock, are the absolute arbiters of truth and value.[6]

Such were the grand prospects conjured by the human imagination on a ground of little evidence. Illusions all!

The Polar Caps of Venus

The next observer of Venus to merit our attention is Franz von Paula Gruithuisen (1774–1852), a staff member of the rural medical school in Munich, where he lectured on physics, chemistry, anthropology and zoology. However, his real interest had been astronomy ever since his imagination was stirred by the Great Comet of 1811, which appeared during Napoleon's march into Russia and is mentioned in Tolstoy's *War and Peace*. Though there is no question that he was a keen observer – among other things, he was the first to make out the delicate Triesnecker rille system on the Moon – he was possessed of an unfortunately overactive imagination. He reported making out a city on the Moon, where today we see only a series of coarse, roughly parallel, ridges. But he outdid himself regarding Venus, whose habitability he took for granted. He was especially concerned with the explanation of the Ashen Light of Venus, a curious phenomenon reported from time to time since 1643, in which the nightside of Venus appears luminous with a faint light, similar to the 'old moon in the young moon's arms' but quite unexpected since Venus has no satellite; even today its reality is doubted, and there has been no generally accepted explanation. Gruithuisen suggested that since one observation of the Ashen Light had been made in 1759, and another in 1806, a difference of 47 terrestrial or 76 venusian years, perhaps the phenomenon was the result of 'a general festival illumination in honour of the ascension of a new emperor to the throne of the planet'. Later he suggested it might simply be due to the burning of large tracts of jungle to make room for farm land.[7]

Variable aspects of the south 'polar cap' of Venus, as recorded by Richard Baum in 1956 with a 12-cm refractor.

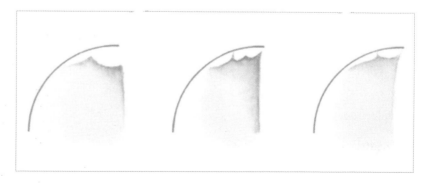

Nevertheless, Gruithuisen made at least one discovery about Venus of enduring value. Observing the crescent Venus in the daylight sky on 29 December 1813 with only a 60-mm refractor, he noticed two very small bright spots near the southern and one bright spot near the northern horn. This was the first observation of the 'polar hoods', which have often been seen since and correspond, as we now know, to cloud swirls generated by the planet's general circulation. As he continued to monitor them, Gruithuisen found the bright spots were seen sometimes at both poles, at other times at one pole alone – again in agreement with subsequent observation.[8]

The Length of Venus's Day

From this series of observations, Gruithuisen was able to determine that the inclination of Venus's axis could not be more than 15°, an improvement on the 10 to 40° Schroeter had published based on far more uncertain indications. However, he does not seem to have given an independent estimate of the rotation period. Not until 1841 did Schroeter's old result have any competition, when the Jesuit astronomer Francesco de Vico at the Roman College published a new value of 23h 21m 22s, only three seconds longer than Schroeter's 1793 value. De Vico and his assistants, six observers in all, were able to track the approach of a 'valley surrounded by

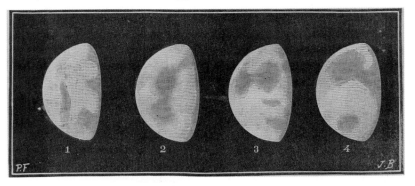

Aspects of Venus according to F. de Vico, 1839, from which he and his colleagues deduced a rotation period of 23h 21m 21.9345s.

The 'Himalayas of Venus', which continued to haunt the lore of the planet from Schroeter's observation of the 'enlightened mountain' until the late 19th century.

mountains like a lunar crater', near the north horn of the waning crescent, which appeared at first as an oblong black spot; afterwards the black spot was bordered with a bright collar, and finally formed a black notch between two bright projections. Subsequently, apparently not wanting to be bested in illusory precision by Schroeter, De Vico revised his period to 23h 21m 21.9345s. 'Never', the Greco-French astronomer Eugène Michel Antoniadi later wrote, 'did scientists . . . indulge more freely in illusions.'[9] E. E. Barnard later chimed in with similar invective. This period, he wrote, 'derived from drawings made a few days apart, with a decimal running into the ten-thousandth of a second . . . ought certainly to be convincing enough, as a smaller subdivision of time would be an insensible quantity and ought never to be stickled for in determining the duration of a planetary day'.[10]

The tendency of successive observers to confirm and refine one another's views was to become a recurring theme in the study of Venus, especially given the elusiveness of its markings and the lack of any clear reference point that can be reliably identified from one apparition to the next. Results given to the ten-thousandth of a second about something subject to such vagaries must stand as one of the most hubristic efforts in the history of science. It illustrates our remarkable ability to see what we expect to see, as well as, perhaps, the tenacious tendency to see Venus as earthlike and possibly inhabited by beings like or unlike ourselves.

71

Schiaparelli''s New Point of Departure

Into the midst of this swelling sea of illusion Giovanni Virginio Schiaparelli, director of the Brera Observatory in Milan, dropped a bombshell in 1878 (the same year he announced the discovery of the 'canals' of Mars). Observing Venus by daylight with the observatory's 22-centimetre refractor in December of the previous year, he had detected a diffuse bright spot at each of Venus's cusps – no doubt the same phenomenon Gruithuisen had seen, though Schiaparelli was evidently unaware of the Munich astronomer's observations. Finding that the bright spots remained fixed in position relative to the terminator, Schiaparelli concluded that the rotation period was as slow as nine months, and probably equal to the planet's year, 225 days. Had he observed only in the twilight periods, as previous observers had done, he might have concluded that the rotation period was the same as Earth; but because he viewed the planet in daylight, he was able to follow developments over several hours, and noted no change. Schiaparelli thus concluded that Venus always kept the same face towards the Sun and the other away, as the Moon does towards Earth. Presumably, this was a result of tidal friction. If Venus's shape was slightly oblong along one axis, solar tides would slow and eventually stop its rotation with the long axis pointed towards the Sun. Mercury, even closer to the Sun and even more affected by solar tides, seemed to represent a similar case, as Schiaparelli announced a few years later.[11]

Inevitably, several astronomers rushed in to confirm the latest trendy result. None did so more provocatively than Percival Lowell, a wealthy Boston businessman turned Far East traveller turned full-time astronomer who, captivated by Schiaparelli's discovery of 'canals' on Mars, established his own observatory at Flagstaff, Arizona, to investigate the possibility that intelligent life might exist on that planet. He had observed Mars with borrowed telescopes in 1894, but in July 1896 installed a permanent replacement telescope

– a 61-centimetre Clark refractor, which after undergoing brief testing on Mercury and Venus in Flagstaff was shipped to Tacubaya, Mexico (a site near the National Observatory), in the quest for better observing conditions for Mars's winter opposition that year. While waiting for Mars, the Mercury and Venus observations continued, and Lowell claimed startling results. Instead of vague and nebulous shadings seen by most astronomers before and since, he reported markings more definite than anything that had been seen since at least Bianchini's time, markings which he described as

> surprisingly distinct; in the matter of contrast, as accentuated, in good seeing, as the markings on the Moon and owing to their character much easier to draw; in the matter of contour, perfectly defined throughout, their edges being well marked and their surfaces well differentiated from one another, some being much darker than others. They are rather lines than spots . . . A large number of them . . . radiate like spokes from a certain centre.[12]

Lowell regarded them as actual features of the venusian surface, seen through the transparent veil of an atmosphere, and could not resist drawing up a map, on which romantic names were conferred from mythology: 'Eros', 'Adonis Regio', 'Acneas Regio' and so on. If real, they definitively proved the 225-day rotation period, and also overturned the whole idea that Venus was shrouded in clouds.

The astronomical world reacted with decided scepticism. Lowell's results were 'entirely at variance with all that has gone before', pronounced the French astronomer Camille Flammarion, while the latter's assistant, Antoniadi, wrote scathingly: 'Forgetting that Venus is decently clad in a dense atmospheric mantle, [Lowell had] covered what [he] called the "surface" of the unfortunate planet with the fashionable canal network, dividing it into clumsy melon slices.'[13]

Lowell's Venus markings have had few
defenders. His biographer William Graves
Hoyt wrote, 'while Lowell could claim that
other observers had seen the canals of Mars,
no other astronomer had, or indeed has, ever
seen anything like the "surprisingly distinct
features" he described on Venus. Not even
his assistants saw them as he did.'[14] Indeed,
it has even been suggested that what Lowell
was mapping were not markings on Venus at
all, but shadows of the blood vessels of his
own retina.[15]

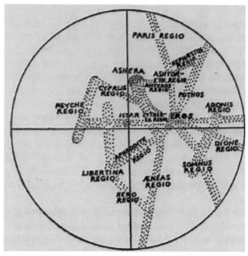

Lowell's map of Venus,
1896–7.

Lowell seems to have been greatly
affected by the criticism, and on returning from Mexico, he
collapsed from nervous exhaustion. For the next four years he
retired from his astronomical work, and not until 1901 had his
condition recovered sufficiently for him to resume work at his
observatory. His first priority was to rebut the harsh criticisms of
his Venus work and to establish the 225-day rotation period that
had seemingly been implied by his observations. He acquired a
spectrograph and an assistant, Vesto M. Slipher, to operate it,
and already in 1903, Slipher found support for the slow rotation
of Venus, though not necessarily the 225-day rotation Lowell had
claimed. Slipher would later put the instrument to much more
significant uses, including making the observations that showed the
large shifts of the nebulae – a foundational observation leading to
the discovery of the expansion of the universe, the most important
discovery ever made at Lowell Observatory.

Lowell would continue to record markings on Venus until his
death in 1916, though his later drawings lack the boldness and
schematic quality of the earliest ones. It is not even obvious that
they represent the same set of markings. In later publications,
he emphasized not their distinctness but their elusiveness, and

An iconic image: Percival Lowell observing Venus by daylight with the 61-cm refractor of Lowell Observatory, 17 October 1914. Though the refractor is impressive, because of the poor daytime 'seeing', it is always necessary to stop down the aperture in order to get the best views of Venus. Lowell typically stopped down to 7 to 10 cm, or even less, for Venus.

now claimed that they were 'hazy, ill-defined, and non-uniform'.[16] In other words, they had come gradually to resemble the usual markings seen by other astronomers.

Of course, Lowell was not the only astronomer interested in Venus, and intriguing results were obtained elsewhere. Streaky markings, though of completely un-Lowellian aspect, were described by a little-known Irish amateur at Lisburn, Ireland, John McHarg. Using refractors of 10- and 12-centimetre aperture during the period from December 1907 to January 1908, McHarg recorded

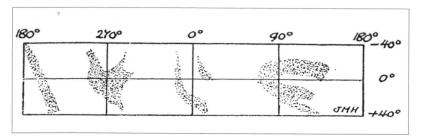

Planisphere of the equatorial regions of Venus, produced by John McHarg at Lisburn, Ireland (now Northern Ireland). The markings, recorded in visual observations from December 1907 and January 1908, strongly resemble those in uv images from Pioneer Venus and other spacecraft, especially the C-shaped feature arbitrarily sited at longitude 90°.

four sets of patches on the disc, and concluded for a short rotation of not greater than 23h 28m. He went so far as to construct a planisphere, published in the *Gazette Astronomique* in 1908.[17] Was he recording genuine markings? We cannot be sure, but at least here we have something that looks rather modern. Indeed, McHarg's planisphere can be usefully compared to ultraviolet maps such as those compiled by Henri Camichel and Pierre Guérin at Pic du Midi Observatory, and Pioneer Venus spacecraft images, about which we will have more to say presently.

It would be tedious, and serve no real purpose, to describe all the visual observations of the planet, or all the rotation periods proposed on the basis of them, which fell mostly into two groups, either the 24-hour period first proposed by Schroeter or the 225-day period of Schiaparelli and Lowell. Typical of the lot is the following observation by Frank E. Seagrave, who, with a 21-centimetre refractor at his private observatory at North Scituate, Rhode Island, on 23 April 1919 examined the planet when it was near the meridian in the daylight sky. He enjoyed a splendid view, and claimed never to have seen the planet so well in 40 or 45 years of prior effort:

> For fully ten minutes the planet was perfectly steady and sharply defined and I am very sure that I could see faint traces of spokelike markings near the centre of the disc. The markings were very much like those seen and described many years ago by the late Dr. Percival Lowell.[18]

Though visual observations of Venus continued to be made, they proved of limited value. Increasingly, the real breakthroughs came about as astronomers began to scrutinize Venus's markings in regions of the spectrum invisible to the human eye, the shorter-wavelength ultraviolet (<380 nm) and the longer-wavelength infrared (>700 nm), both of which became accessible to investigators at about the same time, the mid-1920s.

The Temperatures of the Planet

A word of explanation: the first thing about a planet shining brightly in the night sky – so basic we tend to forget about it – is that what we are seeing is actually reflected sunlight. The light we see belongs to the visible spectrum (380 to 700 nm), and if analysed in the spectroscope, certain wavelengths may be missing due to absorption by compounds in the atmosphere, and hence the identity of these compounds may be discerned. In the spectrum beginning just beyond the red end of the visible range lie the near-infrared (700 nm–2500 nm), and 'thermal emission' regions (3,000 to 25,000 nm), where we are seeing not reflected sunlight from clouds or a planet's surface but sunlight absorbed, stirring kinetic motions of the particles making up the surface or atmosphere, and then being converted from kinetic energy to longer-wavelength electromagnetic energy on being re-emitted again. This energy is partly captured in the vibrations of certain molecules (for example, carbon dioxide, water vapour) and contributes to the warming associated with the greenhouse effect. Given that Earth's own atmosphere contains significant amounts of carbon dioxide and water vapour, a portion of the infrared radiation from a planet, say Venus, will be absorbed, but 'windows' exist in which the absorption by Earth's atmosphere is either fragmented (for instance, between about 2,000 and 5,000 nm) or minimal (between about 5,000 nm and 14,000 nm, corresponding to a gap in the absorption spectrum of water vapour). Near-infrared wavelengths can be accessed using lead sulphide cells (as described later), while thermal emission ('heat') was already being accessed in the nineteenth century by astronomers using special devices such as bolometers and thermocouples. Thus, as far back as the 1870s, Laurence Parsons, the 4th Earl of Rosse, detected heat from the lunar surface, and made an approximate calculation of its temperature.

Since the planets are so much further away than the Moon, and even Venus is at least a hundred times as far, the detection of thermal emission from planetary surfaces requires much more sensitive devices than that used by Rosse. Not until the 1920s did such become available, devised by William W. Coblentz of the radiometry branch of the National Bureau of Standards in Washington, DC. Coblentz teamed up with Carl Otto Lampland of the Lowell Observatory to focus the light and heat emanating from a planetary image onto a tiny thermocouple junction in a glass vacuum tube to produce a small electric current. The electric current produced was amplified and measured in an ironclad galvanometer. By interposing various filter screens or a 1-centimetre-thick water cell that cut off radiation transmission beyond certain wavelengths, the spectrum could be parsed into its visible and infrared components, leading to a calculated temperature of the emitting surface. In this case, the emitting surface was actually Venus's upper cloud deck, where the temperature was very cold.

Making the measurements with Coblentz's thermocouple was a two-person job. As Coblentz later recalled:

Lampland usually operated the telescope and I made the galvanometer readings in a little underground room adjoining the great dome housing the [1.02 m] reflector . . . To the casual visitor . . . who wandered by at night, it sounded monotonous, as we called the signals ('On', 'On'; 'Off', 'Off') as Lampland set the thermocouple receiver 'on' [a] star or planetary image, after which I read the galvanometer deflection, and called 'off' as I recorded measurements.[19]

The studies of Venus became frenetic during the morning apparition of 1924–5. Lampland, after Vesto Slipher the oldest member of the Lowell staff and fiercely loyal to the observatory's founder even in death, expected a planet locked in synchronous rotation relative to the Sun, with (as Lowell had written) 'one face baked for countless aeons, and still baking, backed by one chilled by everlasting night', in which 'indraughts [of air] from the cold to the hot side would blow with tremendous power.'[20] However, a surprise was in store.

Coblentz recalled:

> The first morning when Lampland set the thermocouple on the dark, unilluminated cusps of Venus, not thinking of the implications involved, I called to him that 'a lot of heat' was radiated from the dark portions of the planet, and that it was of different intensities from the tips of the two cusps. Jumping down from the observing platform he came into the galvanometer room, saying that he thought I must be mistaken. Telling him to make the measurements himself, I climbed the ladder to the great reflecting telescope and made the thermocouple settings on the image of the planet, while he read the galvanometer scale – and verified the measurements. He then told me how it might affect the 'P.L. theories.' I admired him for his self-sacrificing loyalty to Lowell.

Thus Lampland and Coblentz were able to show that the measured heat from the day and night sides of Venus was nearly the same, corresponding to a temperature for the upper clouds of about −30 to −40°C. As for the rotation, Vesto Slipher's spectrograms had shown that the rotation was long, though not as long as 225 days, as advocated by Schiaparelli and Lowell. On the other hand, since the temperature changed little between the light and dark parts of the disc, it could not be too slow either, though it had to be longer than 24 hours. Lampland and Coblentz declined to offer a specific figure. Nevertheless, their work was a decided advance. At a blow they had eliminated most of the rotation periods published up to that time (which were bimodally distributed at 24 hours and 225 days).[21]

Ross's Revelation

Meanwhile, the first studies of Venus at the short end of the spectrum, the ultraviolet (wavelengths below about 380 nm), were also carried out during the 1920s, using photography through colour filters.

The first visible-light photographs of Venus had been taken in 1911 by Ferdinand Quénisset at Juvisy Observatory 18 km southeast of Paris, and showed very little contrasts or detail. The earliest photographs using coloured filters were undertaken in 1924 by W. H. Wright at Lick Observatory. However, the real pioneer was Yerkes Observatory astronomer Frank Elmore Ross, who will always be remembered for his discovery of the ultraviolet markings on Venus, one of the most important contributions ever made to the study of the planet.

Born in San Francisco in 1874 and receiving his doctorate at the University of California, Ross started his career as an astronomer pursuing traditional activities such as calculating the orbits of newly discovered satellites of Jupiter and Saturn. However, in 1915 he was hired as a physicist by Eastman Kodak in Rochester, New York, where he specialized in developing colour photographic emulsions and filters. Early photographic materials (silver halide emulsions), including those used in astronomical research, were only sensitive to ultraviolet, violet, blue and green. Thus reddish objects, such as the stars Antares and Betelgeuse, appear decidedly faint in nineteenth-century plates. As early as the 1870s it was found that the

Frank Elmore Ross, c. 1925.

addition of small amounts of certain aniline dyes to a silver halide emulsion could add to sensitivity to colours absorbed by the dyes; chlorophyll, for instance, was found to be a good sensitizer for red. Just before the First World War, such dyes were found that sensitized plates to the near-infrared between 700 and 1,000 nm, and during and just after the war these were rapidly deployed for spectroscopy and haze penetration in aerial photography.

After leaving Eastman Kodak in 1924, Ross accepted a position as an astronomer at Yerkes Observatory, and in 1926–7 was on leave at Mount Wilson, where he obtained photographs of Mars with coloured filters and, at the very favourable eastern elongation of June and July 1927, Venus, using the observatory's 1.52-metre and 2.52-metre reflectors.

Ross's discoveries were somewhat serendipitous. He was not particularly interested in what Venus looked like in the ultraviolet. Instead, his interest was more in how Venus would appear in the infrared. He knew that infrared filters had been used with good effect in terrestrial aerial photography, and assumed that his infrared images would be most likely to penetrate the clouds, and might even show features on the surface. Surprisingly, however, the infrared images proved to be just as bland and featureless as those taken in visible light. Those in blue-violet and blue light showed very weak details. Instead it was the photographs in ultraviolet, taken with a new Wratten 18A filter just released by Eastman Kodak (transmission 300 to 400 nm with a maximum at 360–70 nm) that showed details of bright clouds at what were presumed to be the poles, and dark areas, generally in the form of bands and streaks, in the equatorial and middle latitudes running roughly perpendicular to the terminator. As to what they might be, Ross reached the tentative conclusion that they might represent 'variations in structure of the thin layer of cirrus clouds which overlie the dense yellow lower atmosphere, due undoubtedly to violent disturbances originating far below, perhaps near the surface itself'.[22] This was little more than an educated guess, and even today, the identity of the ultraviolet absorbers remains unknown and constitutes one of the most intriguing still-unresolved mysteries of the planet. As for the rotation period, the testimony of the photographs was hardly clear: the dark bands appeared to be 'belts' indicating rotation, and they appeared to change rapidly from one day to the next. The most Ross could say, like Coblentz and Lampland, and for many of the same reasons, was that the rotation could neither be too short nor too long, and so, in a spirit of compromise, Ross settled on thirty days based on the data he had.

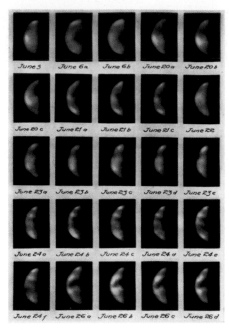

Ultraviolet images by Frank E. Ross from June 1927, using the 1.52- and 2.54-m reflectors at Mt Wilson.

French Connections

During and after the Second World War, French astronomers were undoubtedly the leaders in the study of the planets which, in Audouin Dollfus's words, were still regarded as 'the stepchildren of astronomy' (and largely resigned to amateurs). The level of French activity might seem surprising at first; after all, France had been defeated in 1939, and remained under occupation until 1945. Under the circumstances, it might have been expected that scientific work would have been put aside; in fact, the opposite happened.

The premier site for planetary work was the Pic du Midi Observatory, located at an elevation of 2,877 m on the summit of the Pic du Midi de Bigorre in the French Pyrenees. At the beginning of the war, the large dome of the observatory's 30-centimetre

Pic du Midi Observatory, elevation 2,877 m on top of the Pic du Midi de Bigorre in the French Pyrenees, as it appeared in November 2007.

refractor, which had been used little or not at all during winter months because of difficulties caused by snow and frost, was put in order and employed by Bernard Lyot, Henri Camichel and Marcel Gentili to obtain excellent photographs of planets. Then, in 1943, a new 60-centimetre refractor became operational. It had a very long focal length of 18.24 m (f = 30.4), accommodated within an 8-metre dome by folding the light beam using two flat mirrors. These instruments were specifically designed for planetary work, and could be employed in seeing conditions that were at times nearly perfect. The weather at the summit can change rapidly, and crosswinds perpendicular to the direction the telescope is pointed at can be a problem, since the laminar airflow which gives good seeing is broken into eddies at the edge of the dome shutter

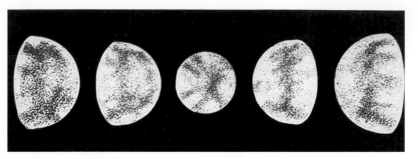

Audouin Dollfus's drawings of Venus, from Pic du Midi. A yellow filter was used, and each drawing is a composite of several in order, it was hoped, that the effects of atmospheric veils would be reduced and the permanent features of the surface revealed. From left to right, these drawings were made on 10 August 1953, 28 August 1953, 7 October 1950, 13 February 1948 and 11 March 1948.

– according to Camichel, who carefully studied conditions at the Pic over many years. But for at least a few nights a month the seeing is magnificent, allowing the 60-centimetre refractor to attain its full theoretical resolution of 0.2″.

In 1946 the observatory staff was joined by Audouin Dollfus, Lyot's PhD student, who went on to become one of the most prolific planetary astronomers of the second half of the twentieth century. He made a long series of visual observations of Venus with the 60-centimetre refractor between 1948 and 1953, in which he emphasized the value of using higher magnifications than usual, pointing out that low magnifications tended to produce spurious details. On Venus, he found the best results with magnifications producing an image some six times larger than the apparent diameter of the Moon as seen with the naked eye. (For a disc 12″ across, corresponding to the fully illuminated planet near superior conjunction, this would correspond to a magnification of 900 ×). In 1955 Dollfus published his results. Instead of showing the planet as it appeared in any one observation, he produced composites obtained by transferring eight to ten drawings made at very close intervals at roughly the same phase to a transparency, and then stacking them. By this process, highly variable details were suppressed and the more permanent aspects emphasized. Dollfus himself believed at the time that his observations had penetrated to a lower layer below the variable upper cloud deck to 'what may be the surface of the planet or a cloud layer associated with the surface topography and having a certain degree of permanence'.[23] What that surface might be was still anyone's guess, and speculations included a watery ocean covering the planet from pole to pole, Saharan-type deserts, or jungles like those that grew on Earth during the Carboniferous period. Dollfus himself did not offer an opinion.

The Amateur's Hour

The French were also the first to follow up on the breakthrough in ultraviolet photography achieved by Ross in the 1920s. However, surprisingly it was not one of the highly trained professionals such as Dollfus who did so, but an amateur astronomer, Charles Boyer.[24]

Born in Toulouse in 1911, Boyer trained as a lawyer, but also at an early age became a ham radio enthusiast, and a mutual enthusiasm for the hobby formed the basis of a lifelong friendship with the professional astronomer Henri Camichel. Before the outbreak of the Second World War, Camichel managed to kindle Boyer's interest in astronomy, and they remained in contact after the war as Boyer embarked on a career in judiciary service in French equatorial Africa, first as chief magistrate at Dahomey (now Cotonou in Benin), then, between 1955 and 1963, as President of the Bench at Brazzaville, the capital of the French colony of Equatorial Africa (from 1960, the independent Republic of the Congo).

At this location only 4° south of the equator, the planets were favourably placed for observation high in the sky. Realizing his opportunity, Boyer built a 25-centimetre Newtonian reflector. It was optically excellent – the mirror had been fashioned by the renowned French optician Jean Texereau – but was set up on a rather primitive altazimuth mounting, and so was ill-suited to making photographs of the planets that required exposures several seconds in duration. But the resourceful Boyer cobbled together a device for moving his camera across the focal plane of his telescope at the proper rate using parts from a Meccano set. Boyer asked Camichel to suggest observing projects. It happened that Camichel was photographing Venus in the ultraviolet from the Pic at the time, and he proposed that his friend have a go at it as well.

In August and September 1957, when the air was unusually dry at Brazzaville, Boyer began to take photographs of Venus, then well placed in the evening sky. He employed Kodak Micro-File film, a high-contrast, fine-grained emulsion that is painfully slow by today's standards. Lacking a proper ultraviolet filter, he made do with a blue-violet Wratten 34 filter that transmitted wavelengths shorter than 450 nm. The images were small and aesthetically unappealing, but he was able to detect the return of the same dusky region to the terminator at intervals of about four days. After five returns between 28 August and 16 September, he alerted Camichel to his findings, and on examining his own images, Camichel too found evidence of a four-day period. Boyer continued the Brazzaville observing campaign until 1960. By then, he and Camichel had come to regard the four-day rotation of the upper atmosphere of Venus as 'completely uncontestable'.

Already in 1957, Boyer had taken the precaution of depositing a sealed envelope describing his discovery with the French Academy of Sciences. The tiny images on which the result was asserted were ambiguous, and not everyone could see the pattern. Dollfus recounted in an interview with one of the authors (W. S.) in 1992: 'I examined the images carefully. They did not seem to me completely convincing at the time.' But with Camichel's unflagging support, Boyer persevered. The four-day rotation became his *idée fixe*, and in fact he would do no other astronomical work of importance before his death in 1989. His first published article, co-authored by Camichel, appeared in the popular magazine *L'Astronomie* in 1960, followed by papers in the prestigious journals *Annales d'Astrophysique* and the *Comptes Rendus de l'Académie des Sciences*. They failed to attract much attention.

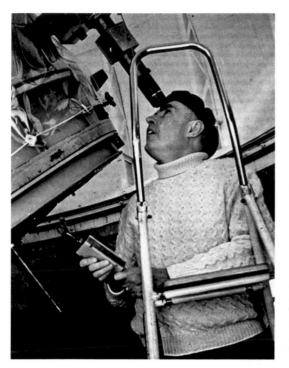

Charles Boyer observing at the 1.1-metre Cassegrain at Pic du Midi. Boyer made one of the most important discoveries ever made by an amateur astronomer – the four day rotation of the upper atmosphere of Venus.

In the meantime, in 1962, radio astronomers bounced radar impulses off Venus's solid surface, and revealed something completely unexpected: a 243-day retrograde rotation of the planet on its axis. This finding seemed utterly incompatible with a four-day rotation rate of the planet's upper atmosphere. How could Venus's cloudtops rotate sixty times faster than the underlying surface? Shortly after the announcement of the radar results, Boyer and Camichel submitted a paper on the four-day rotation to the leading international journal of planetary science, *Icarus*. One of the journal's referees, a strong-willed young Harvard astronomer named Carl Sagan, then making a name for himself as a leading authority on the atmosphere of Venus, rejected it on the grounds that the four-day rotation was 'theoretically impossible', and showed 'how foolish the work of the inexperienced amateur can be'.

That same year, Dollfus and his colleagues Jean Rösch and Jean-Claude Pecker persuaded the International Astronomical Union (IAU) to organize a programme of ultraviolet photography of Venus, in which several stations at different longitudes around the world participated, including New Mexico State University and Table Mountain in California. Following his retirement and return from Africa in 1963, Boyer himself took part, frequently working beside his professional colleagues at the Pic. He and Pierre Guérin succeeded in obtaining scores of ultraviolet photographs of the planet with the Pic's 1.1-metre Cassegrain reflector, installed with NASA funding in 1963. They were able to continuously follow the

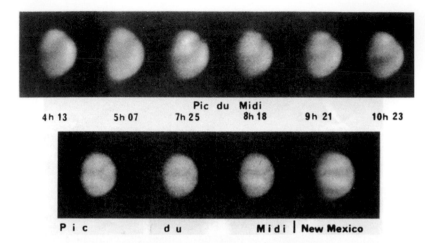

Pic du Midi

4h 13   5h 07   7h 25   8h 18   9h 21   10h 23

P i c   d u   M i d i | New Mexico

Sequences of Venus images in the uv by Charles Boyer with the 1.1-metre Cassegrain at Pic du Midi, showing shifts of uv markings in the clouds due to the atmospheric circulation round the globe in four days. The upper sequence covers an interval of 6 hours on 9 July 1966, the lower 8 and a half hours on 25 July 1966. The last image is from B. A. Smith, New Mexico.

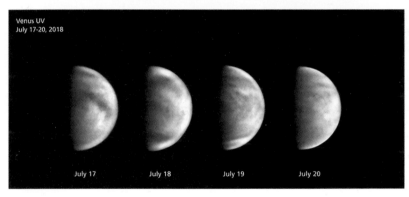

Venus UV
July 17-20, 2018

July 17   July 18   July 19   July 20

Recent images of Venus in the uv by Sebastian Voltmer, 17, 18, 19 and 20 July 2018, with a 31.75-cm Newtonian at ias Observatory in Gamsberg, Namibia.

motions of the ultraviolet markings from east to west for periods of up to six hours. These observations indicated wind speeds of about 83 m/s in the early morning near the terminator. As the markings approached the subsolar point, their drift accelerated, reaching velocities of 122 m/s by mid-afternoon. They also produced a planisphere, on which the now-famous Y- and ψ-shaped markings figure prominently. It looks at least vaguely similar to that produced on the basis of visual observations by McHarg in 1908.

Confirmations

In 1964, another French astronomer, Bernard Guinot, employed
a sensitive technique known as interference spectroscopy to
determine the radial velocities of various points on the limb of
Venus. He found evidence that the upper atmosphere was indeed
circulating with a period of 4.3 days, though final, irrefutable proof
of this awaited ultraviolet imagery from the Mariner 10 spacecraft
upon its encounter with Venus in February 1974. When these
images were combined into a film sequence, the four-day retrograde
rotation of the upper atmosphere was dramatically confirmed.
Dollfus later recalled that when he showed Boyer a copy of this
film, the latter's reaction was one of indifference; it contained no
surprise, since he already knew this result.

So how did Boyer manage to solve one of planetary astronomy's
oldest and most enduring mysteries – a mystery that had defeated
the best efforts of generations of astronomers dating back to the
time of the elder Cassini at the Paris Observatory during the reign
of Louis XIV?

It helped that Boyer was very clear in his purpose and
methodical in his observing programme, and used an excellent

Ultraviolet cloud features on Venus. Above: from Mariner 10.

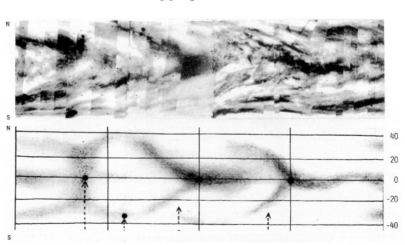

telescope in a favourable climate. But even this does not fully account for his achievement, for three decades earlier Ross had employed far more powerful instruments that recorded far more detail. Here, ironically, Boyer's apparent deficits worked to his advantage. As Dollfus later explained,

> Lack of resolution helped, by making the true picture of what was happening clearer. On images of Venus taken with larger instruments, such as those by Ross and our own at Pic du Midi, there were simply too many details; the sheer profusion of markings confused matters, made it hard to make out the forest for all the trees.

In the end, Boyer the magistrate made judicious use of the rather meagre materials at his disposal. And there was yet one other advantage, noted by Dollfus: 'as an amateur, he had more freedom, as he was not held to the same high standard of rigor that he would have been as a professional.'[25]

Three centuries of intense study had produced a legacy singularly barren of results. The field had been all but abandoned when a persistent and single-minded amateur made one of the last fundamental discoveries in ground-based planetary astronomy. One of Boyer's closest friends, the renowned Romanian planetary photographer Jean Dragesco, summed up the event: 'the case is unique in the history of planetary science.'[26]

With a foothold in a definite result for the venusian rotation, astronomers had finally bid farewell to the long and frustrating prologue which had begun with Giovanni Domenico Cassini in 1666–7. The human eye had proved inadequate to the task of unravelling the riddles of this enigmatic world. Now with Boyer's discovery, which occurred just before the dawn of radar and the spacecraft era, Venus studies expanded at a new and frenetic pace.

FIVE

CHASING SPECTRA

Among all the studies of natural causes Light
chiefly delights the beholder.

LEONARDO DA VINCI, NOTEBOOKS

After centuries of stalemate, in which even the rotation
period remained elusive, by the end of the 1950s Venus was
finally becoming a productive object of scientific investigation.
Though radio astronomy and spacecraft research were about to
revolutionize studies of the planet, photography and spectroscopy,
two technologies first introduced into astronomy during the
mid-nineteenth century, remained important. Spectroscopy, in
particular, was fundamental to addressing the question of whether
or not Venus had water or oxygen, and so to the question that more
than any other has fascinated humans about the other planets
– could it possibly support life? Though spectroscopic studies
of Venus have been mentioned now and again in the preceding
chapters, it seems useful to discuss this in more detail.

The history of spectroscopy begins, in one sense, with Isaac
Newton's experiments with prisms in 1666–7, in which he showed
that white light is composed of all the colours of the rainbow. Other
steps forward came more than a century later, when in 1800 William
Herschel discovered the infrared (thermal) part of the spectrum and
in 1801 Johann Ritter discovered the ultraviolet. However, a major

breakthrough in the development of spectroscopy as a leading tool of astronomical research was due to a Bavarian telescope-maker, Joseph von Fraunhofer, in the early 1820s. While experimenting with the dispersive properties or refractive indices of different kinds of glass in order to attain more perfect achromatic lenses, Fraunhofer invented the modern spectroscope. In 1822 he used it to discover, serendipitously, the series of dark lines in the Sun's spectrum known as Fraunhofer lines. In addition to using a prism to disperse the light, he inserted a slit to narrow the light source in order to make the spectral lines thin. Later, Fraunhofer substituted a diffraction grating for the prism, and so introduced the system that is generally used in modern spectroscopes and spectrographs (the latter using a photographic plate instead of the eye as the detector).

Within a year of his detection of the Fraunhofer lines in the spectrum of the Sun, Fraunhofer trained a spectroscope on Venus, and detected similar, weaker, lines in its spectrum.[1] In both cases, he was only interested in verifying the dispersion or refractive indices of different glasses, and did not investigate them for their own sake. However, he did notice that the lines looked similar no matter how the instruments were rearranged, and so concluded that they could not be artefactual, that is, due to flaws in the instruments or in the experimental design, but rather were an inherent property of solar light. Further, the fact that those of Venus seemed weaker versions of the same lines seen in the Sun's spectrum led him to conclude that the source of the lines in Venus's spectrum was sunlight reflected by clouds in the planet's atmosphere and not its surface.

From Fraunhofer's pioneering start, another three decades were needed before spectroscopy matured into a useful branch of science. In the 1830s, the French positivist philosopher Auguste Comte even went so far as to assert that some things could never be known certainly by any means, among them the chemical composition of celestial bodies. It was one of the most wrong-headed predictions of history as, within a few years, the German chemist Robert Bunsen (inventor of the burner for heating materials to incandescence) teamed up with physicist Gustav Kirchhoff to establish the fundamental laws of spectroscopy, often referred to as Kirchhoff's laws. The first law states that a globe of gas under high pressure, like the Sun, produces a continuous spectrum, or rainbow of colours. The second law maintains that when a substance such as sodium is heated to incandescence, bright lines (emission lines) appear in the spectrum. According to the third law, when the source of a continuous spectrum is viewed behind a cool gas, dark lines (absorption lines) are produced in the spectrum. Kirchhoff asserted that these dark lines are the same as the bright lines produced by emission and hence are produced by the same element. The Fraunhofer lines in the solar spectrum turned out to be absorption lines produced when light from the brilliant surface of the Sun, the photosphere, passes through gas in the cooler solar atmosphere. Hence it is indeed possible – contrary to Comte – to identify a number of elements in the Sun or indeed any star.

The fundamental laws of spectroscopy were published in 1859 and astronomers immediately saw the possibility of applying the spectroscope to the elucidation of the chemical composition of celestial bodies. Venus, known to have an atmosphere, seemed a particularly auspicious target. The incident beam of the sunlight penetrates deep in the atmosphere, but only a fraction of the scattered light is thrown backward towards the observer, so both coming and going, the sunlight effectively passes twice through the atmosphere. The usual absorption lines therefore ought to be effectively doubly enhanced by any constituents present. At first spectra of the airless Moon were used as comparisons, and early spectroscopists, including William Huggins of England, Jules Janssen in France and Hermann Vogel of Germany, reported enhancements in the spectrum of Venus relative to that of the Moon of spectral lines due to water vapour and oxygen, without recognizing that the light had passed through Earth's atmosphere to their spectroscopes.

The observations were difficult, however. This was in the era when photography was still very primitive, consisting of daguerrotypes and calotypes and not yet able to be adapted to astronomical applications, so the observations of the spectrum had to be made visually. This meant holding in memory the strength of the lines in the lunar spectrum for comparison with those in the venusian spectrum. Another complicating factor was that in addition to passing through Venus's atmosphere, the light had to pass through Earth's atmosphere as well. Thus there was no way to avoid the contaminating effect of what were referred to as telluric lines, absorption lines or bands superimposed on the solar spectrum by the oxygen and water vapour of Earth's own atmosphere. The fact that so many of the early spectroscopists nevertheless concluded that the atmospheres of Venus, and Mars as well, showed evidence of the existence of these compounds probably was in large part subjective and attests to a strong human predisposition to believe these other worlds to be earthlike. But in fact because on Venus the amount of water vapour and oxygen is actually very small, it is obvious in retrospect that they could never have been distinguished from Earth atmosphere contributions by means of nineteenth-century instruments and techniques.

The astronomers were hardly naive; they knew very well what they were up against. The bravest attempt to eliminate the contaminating effect of the telluric lines was made by Janssen, a truly remarkable figure of the nineteenth century. He grew up in a cultured family; his father was a celebrated clarinettist and his mother the daughter of a famous architect, and they seem to have been reasonably well off. An accident at the age of eight, apparently caused by his nurse's carelessness, left him permanently lame, and led to his parents educating him entirely at home. At first he entered upon the study of music with his father. Unfortunately, when he

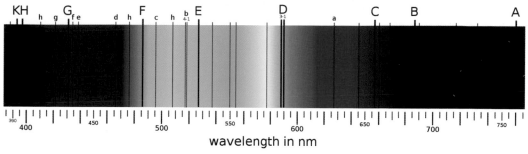

wavelength in nm

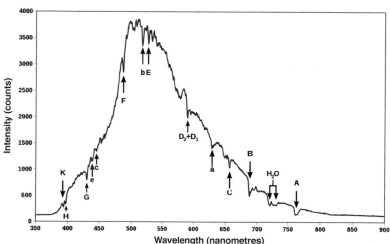

Solar spectrum, showing the positions of the most prominent Fraunhofer lines.

Intensity of the visual spectrum between 350 and 900 nm. The dips correspond to the dark (absorption) lines discovered by Fraunhofer and are designated H, K, G, e, c, F, etc.. The attenuation of the spectrum below about 400 nm is largely due to absorption by ozone in Earth's atmosphere, and above 700 nm to water vapour (of which two bands are indicated by H_2O).

was sixteen, his parents lost their fortune, and Janssen had to begin working for his living. Between 1840 and 1848 he was employed at a bank, and he completed a bachelor's degree at the advanced age of 25. He then attended the University of Paris to gain a teaching licence, and began to work as a substitute teacher. He longed for something more exciting, and turned to science in his early thirties. His first published scientific work was a study of the absorption of thermal radiation in the human eye, in which he demonstrated that the transparency of the optic media of the eye is almost entirely for visual radiation. Thus thermal radiation, being almost completely absorbed, has no deleterious effect on the retina.[2] For this important contribution to ophthalmology, he received a doctorate of science in

1860. This work presaged what would become the defining subject-matter of his career, astronomical spectroscopy. He was among the first to grasp the exciting possibilities opened up by the discovery of Kirchhoff's laws.

He never allowed his lameness to interfere with his expeditions to high mountains (often by sedan chair), balloon ascensions or scientific travels round the globe. In the 1860s he made a careful study from high in the Alps of the absorption of different parts of the solar spectrum by Earth's atmosphere. His conclusion was of lasting importance: most of the so-called telluric lines were produced by water vapour. Determined to get above as much of Earth's atmosphere as possible, in May 1867 he made an intrepid ascent with a spectroscope to the summit of the 3,280-metre volcano Etna in Sicily – an expedition not without danger, as the still-active volcano was smouldering between its last eruption in 1865 and its next in 1868. From these observations he announced the discovery of water vapour in the atmospheres of Mars and Saturn (Venus appears not to have been observed), a result which might have been regarded as somewhat sensational. Janssen himself, however, did not find it very surprising, and declared: 'There are powerful reasons for thinking that life is not the exclusive privilege of our small earth, the younger sister of the great family of planets.'[3]

However, despite these planetary observations, Janssen's main research interest was always the Sun. At the so-called Great Indian Eclipse of 18 August 1868, which he observed at Guntur, Madras State (now in Andhra Pradesh), he noticed hydrogen lines in the Sun's inner atmosphere, the chromosphere, showing, based on Kirchhoff's second law, that the chromosphere must be gaseous. In addition, these lines were so bright that he believed it ought to be possible to observe them in full daylight, and the next day found it to be so. He also found that by proper management of his spectroscope he could study the forms and structures of solar prominences nearly as well as during an eclipse. It was at this same eclipse that he discovered, independently of the British astronomer Sir J. Norman Lockyer, a bright yellow line (587.5 nm) in the solar spectrum that did not correspond to that of any known element on Earth. It was later found to be due to a new element, helium.

The next eclipse visible from parts of Europe was due to occur on 22 December 1870. Unfortunately, this fell during the Prussian siege of Paris. Rather than miss the eclipse, Janssen devised a daring escape from the city by balloon, carrying three telescopes and other equipment as far as the Channel coast. On descending, he continued by more conventional means to Oran, Algeria. Unfortunately, he was clouded out.

Even before the eclipse had passed, astronomers were hurriedly making plans for an even rarer astronomical event than a total eclipse of the Sun. This was the transit of Venus of 9 December 1874. For the first time since Venus had passed off the limb of the Sun in Chappe and Cook's time, the planet would once more track across the face of the Sun. All the great European powers as well as the USA planned multiple well-funded expeditions. Great Britain, then at the height of its imperial sway, took the lead under the direction of the Astronomer Royal at Greenwich, George Biddell Airy, organizing expeditions to Egypt, the Sandwich (Hawaiian) Islands, Rodrigues Island in the western Indian Ocean and Kerguelen Island, which Captain Cook had named the 'Isle of Desolation', in the far southern Indian Ocean. Another British expedition, privately funded by Lord Lindsay (James Ludovic Lindsay, soon-to-be Earl of Crawford and Balcarres), set out for Mauritius. The French planned expeditions to Peking (Beijing), Tokyo, Réunion, Île St Paul, Campbell Island, Nouméa (New Caledonia) and Honolulu. On finding that Réunion was being used by the Dutch and Honolulu by the British, the French instead concentrated their efforts on Île St Paul, Campbell Island and Nouméa. The Germans sent expeditions to Kerguelen Island and Mauritius, and the Americans to Nagasaki and Vladivostok.[4]

Globetrotting Astronomers

As at the eighteenth-century transits, the chief interest at the 1874 transit was again in making observations to determine the precise value of the astronomical unit. The objective was the same, and the basic technique of measuring the contacts from various points across Earth's surface remained fundamentally unaltered. But transport and observational tools had changed beyond recognition since the eighteenth-century transits. Photography had been introduced, and means of travel, including rail and steam-powered ships, had eased global travel to the far points of the globe.

Astronomers were well aware, of course, of the way the 'black drop', whose precise cause was not yet entirely understood, had interposed at the critical moments to spoil the accuracy of the eighteenth-century transit measures. In order to eliminate such effects, they built transit simulators with which to practise observing artificial transits. One of these, used by Charles Wolf of the Paris Observatory, involved a series of lamps and screens set up in the window of a library in the Luxembourg Palace, which could be observed through a telescope at the nearby observatory. They were also aware of apparently subjective effects in which observers differed consistently and systematically by several tenths of a second in measures of phenomena such as the passage of a star across a measuring wire in a telescope field – the so-called personal equation.

Practice would no doubt make better, but it would never make perfect. But what if there were some way of eliminating the vagaries of the 'personality of the eye' altogether, by obtaining an objective and impartial record that could be studied and measured long after the event was over? It seemed to many astronomers, including Janssen, that photography offered such a possibility. With an elegant turn of phrase, Janssen once said, 'I do not hesitate to say that the photographic plate will soon be the actual retina of the scientist.'[5] In the case of the forthcoming transit, he believed that by recording Venus's silhouette against the Sun on a photographic plate, its position could be carefully measured later. And what if an almost continuous series of images could be obtained in rapid succession at the contact points? Here Janssen, hailed by fellow astronomer Hervé Faye as 'the recognized expert on fugitive phenomena', had one of his brainstorms. He devised a mechanism for doing just that. Called the *revolver photographique* (photographic revolver), or usually just 'the Janssen', it consisted of two slotted discs, a large spinning one with twelve slots and a smaller fixed one with a single slot. The latter formed a window that admitted only a small part of the solar image onto a photographic plate. The larger disc was rapidly advanced by a gear mechanism modelled on that in a Colt revolver. Though originally devised only for the specialized purpose of trying to capture the 'fortuitous moments' when the contacts between the planet and the Sun's limb occurred, the photographic revolver had a wider significance; it anticipated (and possibly influenced) the cinematographic process invented by the brothers Louis-Jean and Auguste-Marie Lumière. Indeed, Janssen himself later 'starred' in two early Lumière films made in 1895 that were exhibited to La Société Française de Photographie, of which he was a past president.[6]

As the transit neared, Janssen and his team shipped out for Yokohama (near Tokyo), where the French Transit Commission had decided to locate an administrative base. After battling a typhoon off Hong Kong, during which the crew members encountered all sorts of wrecks and floating bodies and heard reports that more than 1,500 Chinese had disappeared with their sampans, the Janssen expedition finally arrived safely at Yokohama Bay. Rather than set up there, however, Janssen, who was keen to locate the best possible observing site, consulted local experts on the climate predictions for the transit. He found out that the best sites seemed to be at Kobe and Nagasaki. Those at Nagasaki seemed to be most suitable, so Janssen set up his instruments on the ominously named

Jules Janssen, seated in the centre, with his staff and crew bound for Japan. After landing in Yokohama Bay, Janssen's team split up into two detachments, one (including him) observing near Nagasaki and the other at Kobe. As identified by Françoise Launay, those in the photograph are, from left to right: Pierre Marie Arents (standing), Félix Tisserand (seated), Lieutenant Jean Picard (standing), Janssen (seated), Makoto Shimzu, Helmsman Mercier, Francisco Antônio de Almeida Júnior and helmsman Michaut. Janssen's telescope is in the rear, and the revolver camera on the stool at the rear left.

Kompira-yama, the mountain of the god of typhoons, outside the city, but as a hedge he also sent a detachment to Kobe. Fortunately, the weather turned out perfect at both sites, and the day after the transit, Janssen sent a telegram to the president of the Transit Commission in Paris: '10 December. Transit observed at Nagasaki and Kobe, interior contacts obtained. No ligaments [that is, manifestations of the black drop]. Photographed with revolver . . . Venus seen over [solar] corona before transit, demonstrating the existence of the coronal atmosphere. Janssen.'[7]

In addition to deploying the photographic revolver, Janssen used another telescope to place the slit of a spectroscope on the 'aureole' of the planet's atmosphere as the planet stood silhouetted against the Sun's chromosphere. Examining the lines in Venus's spectrum, he claimed to have confirmed the result previously announced on the basis of his Mount Etna observations: there was water vapour in the atmosphere of Venus. As we shall see, he later changed his mind.

The 'philosopher': Jules Janssen, as he appeared in 1895, when he 'starred' in two early Lumière brothers films.

Apart from the deployment of 'Janssens' by the French, the British had their own versions, constructed by the London instrument-maker John Dallmeyer and used in India and at Melbourne.[8] The effort expended to observe the transit was unprecedented, including those using more conventional means. Observations were made from four continents and numerous islands scattered across two oceans. And yet, the results, and the calculation of the astronomical unit, fell short of expectation. The British astronomer David Gill, a member of Lord Lindsay's expedition who observed the transit using a heliometer (a telescope with a split objective lens used to find the angular distances between stars), resigned from Lindsay's employ on returning to Britain. Disenchanted with his experience with the vagaries of determining Venus's exact position relative to the Sun during the transit, Gill decided to concentrate his efforts on a different method of measuring the Sun's distance, using Mars's changing position relative to the stars at a close opposition. In time for the Mars opposition of 1877, he and his wife Isobel travelled to the desolate island of Ascension, in the mid-Atlantic, and using the same heliometer he had used to observe the transit of Venus, worked out a value of the solar parallax of 8.780″, making the Earth-to-Sun distance 149,840,000 km, within 0.2 per cent of the modern value. Even with the help of photography and the 'Janssens', the transit observations produced a scattering of results, of which those produced by the British photographic teams were especially disappointing. Only by discarding the latter do the 1874

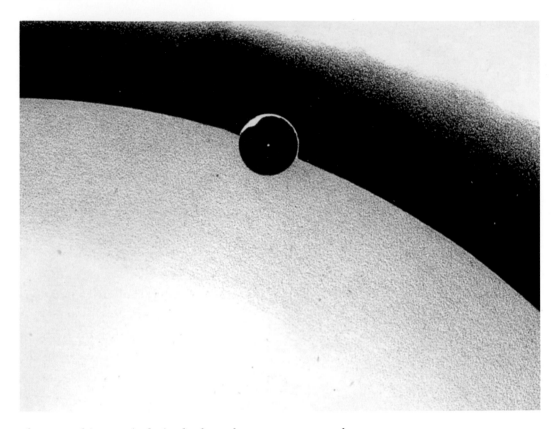

photographic transit-derived values show greater mutual agreement than the black-drop compromised visual timings of the eighteenth century. As Gill recognized, the geometric methods of Halley and Delisle had become passé, and henceforth the trigonometric approach to the Sun's distance would use Mars, as Gill did – or better yet, one of the asteroids – as an intermediary. Though this realization had already begun to sink in well before the next transit, that of 6 December 1882, institutional inertia meant that the planned armada of transit expeditions continued to a grand finale.

On 8 December 1882, the *New York Times* disapproved of the extravagance in appropriating taxpayer funds for this purpose, and lambasted the U.S. Naval Observatory for sending expeditions to remote points of the globe when, they argued, in contrast to the

Venus closing in on the end of its transit, 9 December 1874. A brightening known as the 'polar spot', located near the presumed position of the planet's south pole, as recorded by Dr H.G.A. Wright at Sydney, New South Wales, using a 2-cm Browning silvered glass reflector stopped down to 13 cm.

Edmond-Louis Depain, *Transit of Venus*, 1882. Painted ceiling at the Paris Observatory commemorating the transit of Venus of December 1882. Venus, accompanied by a putto, is seen flying towards Helios and his horse-drawn chariot. This sort of thing was still in vogue in the 19th century, but is less likely to appeal to modern tastes.

1874 transit, that of 1882 could just as well be observed from within the continental USA, with both the beginning and end being visible. They wrote derisively:

> People have become so accustomed to accept the assertion of astronomers without question that no one has ventured to ask how the passage of Venus across the sun's disc enables any one to calculate the distance of the latter body from the sun . . . It has finally become evident to all thinking people that as long as astronomers can induce Governments to send them to China and Peru the necessity of observing transits of Venus will never come to an end, and the distance of the sun from the earth will never be definitely settled.

British, French, Dutch, German, Belgian, Spanish, Brazilian and Argentine expeditions were also sent out. The results justified the *New York Times*'s scepticism and showed that, in terms of precision in determining the value of the astronomical unit, the transit method was indeed obsolete. (As a reminder, Kepler's harmonic law relates the period of revolution to the distance of the planet from the Sun. This allows a scale model to be worked out, but to get the absolute scale in terms of kilometres or miles, the distance between any two bodies needs to be measured. By measuring the distance from Earth to Venus – or Mars – the distance to the Sun, or astronomical unit, follows directly.) The latest measurements have been made not by the complicated triangulation of Halley's or Delisle's methods, but by noting the time lag between sending radio signals to, and receiving them back from, the surfaces of Venus or Mars. Radio signals travel at the speed of light: 299,792 km/s. The resulting distances are accurate to within a few metres. Since 2012, the solar parallax has been defined (rather than measured) by the International Astronomical Union as 8.794148″, and the astronomical unit defined as 149,597,870 km (92,955,807 international mi.). Since this is a definition and not a specific measurement, it is likely to be the last word.

On the other hand, the 6 December 1882 transit was not without significance. It was the first transit for the 'common man', accessible to anyone equipped with a shard of smoked glass. Those who made the effort were rewarded by the sight of a speck that was, according to the *New York Times*, 'so small that it required some time of close application to the glass before it was recognizable. The dark spot appeared no larger than a small-sized dried pea.'[9] In itself, it was hardly much of a spectacle, yet it was surely something to have seen that which would never be seen again by anyone living. The same was true of the even more widely observed transits occurring in our own lifetimes, those of 8 June 2004 and 6 June 2012. The next ones do not occur until 11 December 2117 and 8 December 2125.

Janssen also observed the transit of Venus of 1882. He
completely ignored the international campaign to determine the
solar parallax, and instead attempted to make physical observations
of Venus, using a spectroscope to analyse the composition of the
atmosphere of the planet. His goal was to determine 'the presence
or absence, in that atmosphere, of that aqueous substance which,
on Earth, plays such a great role in all the phenomena that are
related to the development of life'.[10] Rather than remain in Paris,
he travelled to Oran, Algeria, the scene of his unsuccessful 1870
eclipse adventure. There the Sun would stand higher in the sky
than in France at the moment the transit got underway. He did
not succeed in his initial attempts to detect lines due to water
vapour in the aureole of Venus, but unwilling as yet to give up, he
extended his stay in the high-plateau region for a month. The seeing

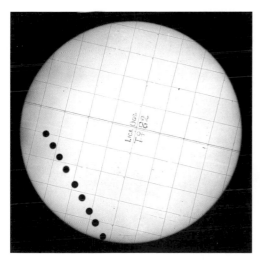

A montage created by
Anthony Misch using David
Peck Todd's plates taken
on Mt Hamilton, the future
home of Lick Observatory,
on 6 December 1882.

Children peering at that
which would never be seen
again (in their lifetimes),
Venus in transit across the
Sun, on 6 December 1882.
From *Harper's Weekly*,
28 April 1883. Their method
of observation is not
recommended.

conditions in Oran were excellent. He noted both the extraordinary 'brilliance' of the planet, enabling high-dispersion and high-quality spectroscopes to be used, and the extreme dryness of the site, which minimized the interference from the telluric lines. And yet, in the end, he had to admit that 'when one . . . eliminates the influence of the terrestrial atmosphere, the optical characteristics of water vapour in the spectrum of Venus are very weak . . . This does not mean, to my mind, that this element is absent in Venus.'[11] Thus, as Janssen made clear, the quest for water vapour, which did not require a transit, would continue.

As had been evident by 1882, the quest for the astronomical unit did not require a transit either. The Halley and Delisle methods had been superseded, though they had not been without value, and had marked the beginning of the era of international scientific expeditions. It is sobering, by the way, to note that not every generation can see a transit of Venus. The next transits after 1882 would be those of 2004 and 2012, only the sixth and seventh

The 'Venus Twilight Experiment' was a worldwide effort to deploy coronagraphs to stations around the globe to image the aureole of Venus during the transit of 6 June 2012: Kazakhstan, where the observations were spoiled by clouds; Longyearbyen Svalbard; Udaipar, India, with João Retrê and Pedro Machado (whose visit for the transit was coordinated with Udaipur Solar Observatory by Sanjay Limaye) with coronagraph aboard elephant.

since that observed by Horrocks and Crabtree in 1639. In 1882 they seemed very far away indeed, though they have now come and gone, observed by more people than any other transit in history, probably hundreds of millions. Those who missed them will have to wait until 2117 for another opportunity.

Though most observers of the transits were doubtless motivated merely by the pleasure in witnessing a rare astronomical event, the twenty-first-century transits were not without scientific value. One observing campaign, the 'Venus Twilight Experiment', involved sending an identical series of specially designed telescopes to observing stations around the world. These telescopes were of a type originally designed for observing the solar corona at times other than during a total solar eclipse, and hence referred to as coronagraphs. In this case, they were perhaps better known as cytherographs, since they were used to study the aureole as Venus passed off the edge of the Sun. As already known from spacecraft data, the vertical structure of Venus's atmosphere shows decreasing temperatures from the cloud level to the surface. Above the clouds the rate of change of temperature with altitude is lower, even close to zero at some latitudes between 70 and 80 km. Since the thermal structure impacts atmospheric density, it also affects refraction of light. The aureole consists of light refracted in the mesosphere. Data from the 'Venus Twilight Experiment' stations favoured by good weather, in combination with that from the Solar Dynamic Observatory and Venus Express spacecraft, allowed the vertical structure of the mesosphere to be worked out in detail, and showed evidence of short timescale variability and latitudinal differences in the layer above the cloud decks. For instance, the bright 'polar arc' seen by transit observers since 1874 was found to correspond to refraction of sunlight in the planet's 'cold collar region', a bulge of cold air at 60–80° latitude extending vertically up to 75 km. Not only did the observations reveal important results concerning Venus itself, they served a secondary purpose as a dry-run for future astronomers'

attempts to analyse phenomena exhibited during the transits of
distant exoplanets circling other stars.

A First with the Spectrograph: Not Water Vapour but Carbon Dioxide

Further observations dispensed with visual efforts, which were
hampered by subjective effects, and instead relied on photography.
Janssen had proved prophetic; by the early twentieth century,
the photographic plate had in most areas of astronomical
investigation become the 'actual retina of the scientist'. This was
true in spectroscopy, where the spectroscope now became the
spectrograph.

As mentioned earlier, Vesto Slipher at Lowell Observatory was
investigating Venus's rotation with the observatory spectrograph as
early as 1903. In addition, his employer, Percival Lowell, suggested
a promising new method for investigating the compositions
of planetary atmospheres. As we have seen, this is all rather
complicated, because of the difficulty of disentangling water
vapour and oxygen in Venus's atmosphere from that in Earth's.
Lowell suggested, however, that this might be done by means of
the well-known Doppler effect. The principle is this: when Venus is
approaching Earth, the telluric lines (those due to the Earth) remain
steadfastly in place, but the venusian ones are shifted to the blue,
while when Venus is receding from Earth, the venusian ones are
shifted to the red. Thus the lines due to water vapour, oxygen or any
other element ought to appear to be doubled, or at least somewhat
broadened, and those belonging to Venus distinguished from those
due to the Earth.

In principle, the method is quite straightforward, but it is not so
easy to put into practice. A first attempt to do so was made by Slipher
at Lowell in 1921, then by Seth B. Nicholson and Charles E. St John at
Mount Wilson the following year. Neither oxygen nor water vapour
were detected. In 1932 Walter Sydney Adams and Theodore Dunham

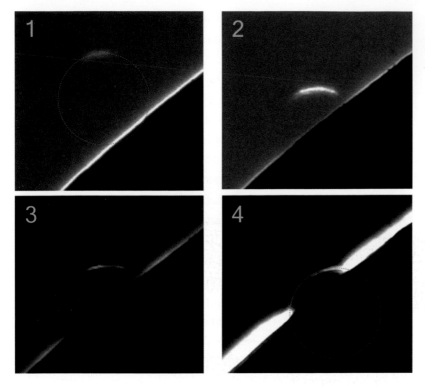

The aureole, imaged on 8 June 2012. These CCD images were taken by Paolo Tanga with a Venus Twilight Experiment coronagraph, one of an identical series of specially designed telescopes sent to various locations around the world which were used to study Venus's aureole as it appeared just off the limb of the Sun at the beginning and end of the transit of 8 June 2012. This one was set up at Lowell Observatory in Flagstaff, Arizona. Note the bright 'polar arc' in panel 1, first observed by visual observers in New South Wales at the 1874 transit.

Jr, using photographic plates with improved sensitivity in the infrared part of the spectrum on the 2.54-metre reflector (the largest telescope in the world at the time), also failed to detect oxygen or water vapour. Instead they unexpectedly discovered absorption bands in the infrared. They could not at first be identified. Adams and Dunham later showed them to be due to carbon dioxide.[12]

The story takes an interesting twist at this point. In 1937 Arthur Adel, trained in infrared spectroscopy in the lab of Harrison M. Randall at the University of Michigan, was on the staff at Lowell Observatory. He had difficulty getting unpublished data from the older astronomers there, including Slipher, but while waiting for the observatory's instrument-maker to finish working on his equipment he was allowed to use the 61-centimetre refractor (the one Percival Lowell had mainly worked with), and Slipher allowed him to put the spectrograph on the telescope. According to Adel's later account,

> I immediately turned it onto Venus. I used new photographic plates . . . [that] had to be hypersensitized with ammonia and one thing and another, and then they became sensitive out to about a micron or so [the near-infrared] . . . So with this equipment, the first thing I did was pick up the bands, the carbon dioxide bands in the spectrum of Venus that Adams and Dunham had discovered . . . which meant that V. M. Slipher could have done that. He could have made the discovery of carbon dioxide in Venus at the Lowell Observatory. Well, I didn't pay any attention to that . . . I was naive enough not to realize how offended he would be if this was called to the attention of [anybody]. Dunham came for a visit to Lowell in '37 . . . and I was foolish enough to mention it. I did it out of enthusiasm . . . and I showed him the plates. V. M. was present, unfortunately, when I told Dunham this. After that, I was not permitted the use of any of the telescopes on the Hill.[13]

Spectrograms by V. M. Slipher at Lowell Observatory. In this series, spectra of Venus are compared with those of the Sun, in the attempt to discover enhancement of absorption lines due to the presence of water vapour, oxygen and so on.

Despite ongoing friction with the older scientists at Lowell, which was never completely resolved and led to his departure at the outset of the Second World War, Adel continued to do brilliant work. In 1941 he published a paper in which he estimated that the amount of carbon dioxide in the atmosphere of Venus was equal to a

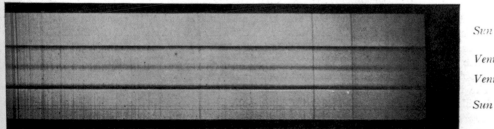

Sun

Venus

Venus

Sun

layer 3.2 km in thickness at a standard atmospheric pressure and temperature, compared with a mere 9 m for Earth.

A Stratospheric Achievement: The Detection of Venusian Water

The first actual detection of venusian water vapour did not take place until the late 1950s, when, just before the first interplanetary probes became feasible, stratospheric balloons allowed instruments to be carried to the edge of space.

Among these exploits were those of the famous astronomer Audouin Dollfus, whose observations at Pic du Midi were described previously. In May 1954, with his father Charles, a celebrated French aeronaut, serving as pilot, Audouin made observations from an altitude of 6,400 m in an attempt to detect the 830 nm water vapour band in the spectrum of Mars. On this occasion, Venus was not observed, but the effort proved a rehearsal for a subsequent ascension on 22 April 1959, when both Venus and Mars were observed. This time Dollfus was alone, acting as both pilot and observer. He was enclosed within a sealed, spherical aluminium–magnesium alloy gondola, and able to operate a 51-centimetre telescope mounted outside. The whole device was lifted aloft by 104 neoprene balloons spaced along a 400-metre length of nylon rope to an altitude of almost 13,000 m. As soon as the observations had been recorded, a radio command released several balloons at the top of the cluster to start the descent. The

The 104-balloon cluster and payload shortly after launch from Villacoublay airport, near Meudon, 22 April 1959. Using a telescope mounted above the gondola, astronomer Audouin Dollfus searched for water vapour on Venus and Mars.

Dollfus emerges triumphantly from the gondola shortly after touchdown.

night landing was uncontrolled, but immediately after touchdown all the balloons were released into the dark sky, leaving the capsule, telescope and astronomer safely on the ground. Despite the valiant effort, Venus's water vapour continued to elude detection – one more proof of the extreme dryness of its atmosphere.[14]

Success came the following November, when two Americans, Malcolm D. Ross serving as pilot and Charles B. Moore as observer, carried a 41-centimetre telescope and a spectrograph to an altitude of 25 km, and thus above most of Earth's atmosphere, in Stratolab IV. The launch took place from the famous Stratobowl in the Black Hills of South Dakota, which had been used for a series of record-breaking balloon flights going back to the mid-1930s. This time unmistakable signs of water vapour were detected, though a quantitative result was very uncertain. It did seem, however, that the amount of water vapour on Venus above the clouds might be 'about the same as lies above high-level clouds on the Earth'.[15]

There matters stood on the eve of the spacecraft era.

SIX

A NEW ERA: RADAR AND SPACECRAFT

A musical warble, the voice of *Mariner 2*, resounded in the hall
[NASA headquarters in Washington, DC] and in millions of
radios and television sets around the world . . . To the world
at large, this warbling tone was a signal that the United States
had moved ahead – reached out to the planets. *Mariner* was
exploring the future, seeking answers to some of the unsolved
questions about the Solar System.

'MARINER: MISSION TO VENUS', THE STAFF OF JET
PROPULSION LABORATORY, CALIFORNIA
INSTITUTE OF TECHNOLOGY, 1962

The modern understanding of Venus came about as a result
of several developments brought about or hastened by the
Second World War. One was the rocket, which though invented
centuries ago, only now became a vehicle capable of reaching
outer space. Another was radar, short for RAdio Detection And
Ranging, a system capable of producing short pulses of radio energy
(1 mm to 100 km wavelength). It was independently developed
by eight nations (the UK, Germany, the USA, the USSR, Japan,
the Netherlands, France and Italy) between 1934 and 1939, and
rapidly amped up for military purposes during the war. A third
was the lead sulphide cell, which produced an electric signal when
exposed to heat. It was far more sensitive than thermocouples, as

used by Coblentz and Lampland for detecting wavelengths longer than were accessible to the human eye, or even the most sensitive photographic emulsions.

At least in primitive form, rockets have been around since the Chinese invented them centuries ago and filled them with another of their inventions: gunpowder. They were developed by the Englishman William Congreve for use during the siege of Copenhagen in 1807, and also played a minor role during what was known in Britain as 'the American War of 1812' (hence the 'rockets' red glare' in the 'Star-Spangled Banner'). The Russian schoolteacher Konstantin Tsiolkovsky published numerous papers on the theory of rocket propulsion beginning in 1902, while just after the First World War, an American physicist, Robert Goddard, published a treatise in which the rocket was proposed as a method of reaching extreme altitudes for upper-atmospheric research. Goddard also suggested that a rocket might even succeed in reaching the Moon, and that such a vehicle, if loaded with flash powder, might be seen from Earth when it crashed on impact. Predictably, Goddard was viewed as a crank, and in a *New York Times* editorial the following year was accused of ignorance of the most elementary principles of physics in imagining that a rocket could work in a vacuum. Undaunted, but continuing to work more secretively, he launched the first liquid-propellant rocket on his Aunt Effie's farm near Auburn, Massachusetts, on 16 March 1926, and continued his rocket experiments in the 1930s from a ranch near Roswell, New Mexico. Meanwhile, others were entering the field, including Hermann Oberth, a Romanian-born schoolteacher whose book *Rakete zu den Planetenräumen* (The Rocket to Interplanetary Space) inspired a group of German rocket enthusiasts, the Verein für Raumschiffahrt, or Society for Space Travel, among whose leading members was Wernher von Braun.

Though Goddard worked alone and was unable to interest American officials in the potential of his rockets, the German army quickly grasped the rocket's military potential, and in 1932 von Braun was called to Kummersdorf to serve as a consultant to the army's new rocket station there. After the Nazis came to power, he was placed in charge of a much larger complex, at Peenemünde in the Baltic, where the deadly v-2s (from *Vergeltungswaffe* 2, or Retribution Weapon 2), the world's first long-range guided ballistic missiles, were developed. Beginning in September 1944, over 3,000 v-2s, produced largely by forced labourers and concentration camp prisoners, were launched against Allied cities such as London, Antwerp and Liège. They were weapons of terror, but although they caused the deaths of an estimated 9,000 civilians and military personnel, as well as 12,000 labourers and concentration camp prisoners, they were unleashed too late to change the course of the war. At the end of the war, von Braun and most of the other German rocket engineers surrendered to the Americans, although a few were captured by the Soviets. An American rocket programme was quickly organized under their leadership. Several captured v-2s were fired from White Sands, New Mexico, and whereas the v-2s were single-stage, a multi-stage rocket, consisting of a v-2 first stage and a smaller rocket known as the WAC Corporal for the second stage, was successfully fired in 1949, reaching an altitude of 400 km.

Radar also advanced rapidly because of its military applications, especially after the development of the cavity magnetron by John Randall and Harry Boot at the University of Birmingham in 1940, which allowed relatively small systems with sub-metre resolution to be widely deployed for the detection of military targets such as aircraft, motorized vehicles and ships. It has been credited as one of the factors leading to victory by the Allies. Just after the end of the war, on 10 January 1946, a team of military and civilian engineers at Fort Monmouth, New Jersey, succeeded in reflecting radar signals off the Moon and receiving them on their return. The signals took 2.5 seconds to make the round trip. This was the first radar detection of a celestial object.

The electromagnetic spectrum, from gamma-rays, X-rays and the ultraviolet at wavelengths shorter than the visual part of the spectrum, to infrared (including thermal IR), microwaves, and radio waves at the long-wavelength end. Note that before the 1950s, the only part of the spectrum of Venus accessible for investigation was the slender sliver between the ultraviolet down to about 300 nm and the thermal IR at around 1,000 nm.

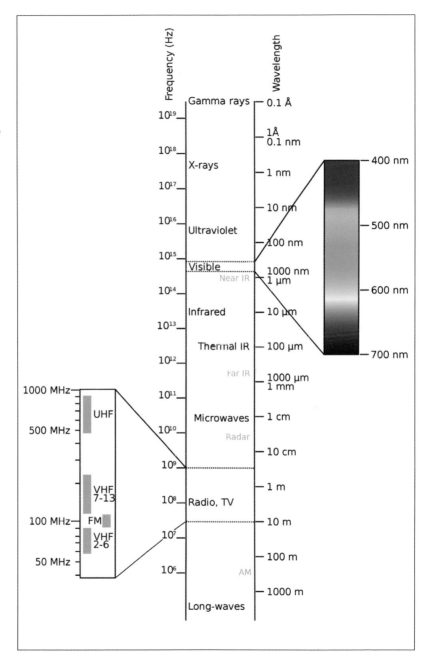

The Germans had also made great strides with lead sulphide detectors, which they had hoped to use to guide v-2s to sources of heat in order to maximize destruction of Allied military and civilian targets. After the war, Gerard Peter Kuiper, of the Manhattan Project's Alsos Mission (Allied personnel sent behind front lines to assess German progress in developing the atomic bomb and other weapons), learned of this work from his interrogations of captured German scientists. On returning to the USA, where Kuiper was on the staff of Yerkes Observatory and the University of Chicago, he discovered from newly classified documents that the Americans led by Robert J. Cashman, a physicist at Northwestern University near Chicago, had also been trying to develop lead sulphide detectors. Given their proximity to one another, they began to collaborate on the use of these detectors for scientific research. This was an important step leading to the development of infrared astronomy of solar system bodies since they allow access to part of the near-infrared spectrum where absorption by Earth's atmosphere is fragmented, as described earlier, and where emission and absorption bands produced by the principal components of our atmosphere, such as nitrogen, oxygen and carbon dioxide, exist as gases, and for more distant bodies, commonly linked with hydrogen, as water ice, methane and ammonia. Unfortunately, we cannot go into more detail about the development of infrared astronomy here, but hopefully enough has been said to orient the reader to discussions of infrared observations of Venus, both ground-based and by spacecraft, in what follows. Meanwhile, for convenient reference, the different regions of the electromagnetic spectrum are summarized in the image in this section.

Microwave Studies of Venus

In addition to these explorations of the infrared part of the electromagnetic spectrum, microwave radiation from Venus

(1 mm to 1 m wavelength) was first detected in 1958, with the 15-metre radio telescope at the Naval Research Laboratory in Washington, DC. These observations provided a rough indication of the surface temperature, which proved to be on the order of 600 kelvin.

In 1960 Carl Sagan, a graduate student under Kuiper at the University of Chicago, and soon to become the most recognizable public spokesperson for planetary exploration, referenced this result in his PhD dissertation, in which he argued: 'It is evident that [this surface temperature] demands a very efficient "greenhouse effect".'[1] We need not say much to introduce the concept of greenhouse warming here, as it has been widely discussed in recent years. Briefly, despite being present in only trace amounts, the so-called greenhouse gases – including carbon dioxide, methane, nitrous oxide and water vapour – are very efficient absorbers of infrared radiation while also being transparent to incoming solar radiation, and they play a crucial role in determining Earth's mean temperature. Indeed, but for the warming these gases produce, Earth's mean global temperature would drop to −20°C (−4°F) and Earth's oceans would freeze over. Since the Industrial Revolution, greenhouse gases have been rising significantly, in large part owing to human activities, and at accelerating rates in recent years. Thus, from March 1958, when Charles Keeling began measuring carbon dioxide levels from Mauna Loa Observatory on the north flank of the Mauna Loa volcano on the Big Island of Hawaii, to November 2018, the annual average carbon dioxide concentration in Earth's atmosphere increased from 315 ppm to 406 ppm, an increase of 2.48 ± 0.26 ppm per year. It had reached 417 ppm by May 2020.[2] Earth's mean temperature has been undergoing a corresponding increase of some 0.6°C per century; a de facto confirmation of the greenhouse warming is now consistent with climate model forecasts.

Venus's 'brightness' in the microwave region indicated temperatures of hundreds of degrees Celsius. Soviet scientists

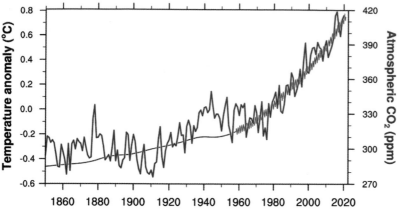

The Keeling curve, plotting atmospheric CO_2 vs change in mean Earth temperature. The dates shown are from 1850 to 2020, with the earlier data approximate. Keeling's own measurements from the top of Mauna Loa in Hawaii began in 1958 (blue line). Squiggliness is due to the approximately 5 ppm variation in carbon dioxide caused by seasonal change in carbon dioxide by the world's land vegetation, most of which lies in the northern hemisphere. Note that the temperature variation tracks very neatly the carbon dioxide concentration.

argued that these temperatures might be ionospheric rather than those at the surface of the planet, but Sagan famously, and correctly, argued that they were indeed those in the lower atmosphere and at the surface. Further, he believed they could only be explained if, in addition to very substantial amounts of carbon dioxide, there was also a significant contribution from water vapour at high altitudes above the surface. Even then, no water vapour was generally believed to exist on the surface itself. He went so far as to attempt to model the warming based on several different assumptions about the planet's rotation rate, which at this point was still unknown. One plausible model assumed an ice crystal cloud layer about 36 km above the surface.

Radar Studies of Venus

As noted, the Moon was 'detected' by radar as early as 1946. Naturally Venus, being always at least 100 times further away, took longer. An initial success claimed during the period around inferior conjunction in 1958 proved to be premature. However, during the next favourable opportunity, the period near inferior conjunction in March–May 1961, no fewer than four teams were successful: William B. Smith's with the Millstone Hill steerable antenna of the MIT

Lincoln Laboratory; Robertson Stevens and Walter K. Victor's with Goldstone antenna in the Mojave desert in California; a Manchester University team with Jodrell Bank radio telescope in England, and Vladimir Kotelnikov's team using the Evpatoria radio telescope in the Crimea. As an immediate result of this work, they were able to calculate a refined value of the astronomical unit (AU), the distance from Earth to the Sun. It was also apparent that the surface of Venus was smoother than that of the Moon.

Just prior to the following inferior conjunction, in November 1962, Smith reanalysed his 1961 data, and noticed that the radar echoes contained two components. One was a time-delayed component, which tells the planet's distance from the Earth. The other was a Doppler-shifted component, which tells the rotation, since higher frequencies than that of the original signal return from the approaching limb of the rotating planet, and lower frequencies from the receding limb. His data seemed to suggest a surprising result, that the direction of the rotation of the solid body of the planet is opposite (retrograde) to the direction of its orbital motion.

This slow retrograde rotation was unexpected at the time, but fully confirmed during the 1962 observing window by Roland L. Carpenter and Richard M. Goldstein using the Goldstone antenna. In addition, Carpenter and Goldstein found a sidereal rotation period, the rotation period relative to the stars, of some 250 days.[3] Later studies using the Goldstone antenna and larger antennae elsewhere, including the giant 305-metre Arecibo radio telescope, had by 1967 further refined this to 243.01 days.[4] The most recent figure is 243.0185 days. These results imply that the solar day on Venus, the interval from one sunrise or sunset to the next, is 117 Earth days. Daytime on Venus thus lasts 58.5 Earth days, followed by a night of equal duration. Notably, the 4.3 Earth-day rotation discovered by Boyer is that of the upper cloud layer; the faster rotation of the atmosphere compared to that of the solid body of

The MIT Lincoln Laboratory team that attempted to bounce radar signals off Venus in 1958. They were unsuccessful in this first attempt, but succeeded in 1961. From the right: Paul E. Green Jr and Robert Price, the experiment's originators; Thomas J. Goblick Jr, Gordon H. Pettengill (who replaced Green as team leader in 1960), computer programmers Roland Silver and William B. Smith, and Leon G. Kraft Jr, an operator of the transmitter. On the blackboard is a block diagram of the Venus radar instrumentation.

the planet constitutes one of the great puzzles about Venus, the atmosphere's 'superrotation'. We will return to this later.

Analysis of the radar echoes allowed not only the rotation to be discovered but the topography to be roughly worked out. This led to the publication of a series of increasingly detailed radar maps made with the Goldstone radio telescope and the interferometer on the 305-metre radio telescope at Arecibo, Puerto Rico. Their resolution was low compared to later spacecraft radar mapping results, but they were astounding achievements for the time, and good enough to show that most of the planet consists of rolling plains, with some elevated areas. These were referred to at the time as 'continental' regions, and are still often referred to as 'continent-like'.

A state-of-the-art radar map made with the Arecibo radio telescope in 1970. From D. B. Campbell et al., 'Radar Interferometric Observations of Venus at 70-Centimeter Wavelength', *Science*, CLXX (1970), p. 1090.

Spacecraft to Venus

At the same time as the radar mapping of Venus was getting underway, the first spacecraft were setting out across interplanetary space. That heady time began which, in retrospect, would seem to be a golden age of solar system exploration, as well as the birth of planetary science as a discipline. Sixty years later, the planetary science has matured, but those who lived through the early days remember each attempt to reach the planets, successful or aspirational, as a stunning feat etched in bold characters.

The v-2 had clearly demonstrated the military potential of the rocket, and at first, such is human nature, this remained the chief interest of the rival superpowers, the USA and what was then the USSR. In July 1955 the USA announced plans to launch the first artificial satellite around Earth. At the time, American technological superiority was taken for granted in the West, and the USA was generally assumed to have the field to itself. In fact, the Soviets were further ahead than anyone knew, led by their brilliant engineer Sergei P. Korolev, whose own experiments in rocketry had begun in the 1930s, and who had survived, though with great damage to his health, being denounced and imprisoned during the Great Purge of the Stalin years. On 4 October 1957, Korolev and his team launched Sputnik 1 from the Soviet launch facility on the desert steppe of Baikonur, near Tyuratam in central Kazakhstan. The Americans launched their own artificial satellite, Explorer 1, on 31 January 1958; it was notable for carrying aloft the instrumentation that first led to the identification of the Van Allen radiation belts surrounding Earth. In 1959, the Explorer 7 carried the first experiment to measure the radiation budget of Earth by Verner E. Suomi. The Soviets continued to maintain an early lead, with successful lunar missions including Lunik 3, which returned the first photographs of the Moon's hitherto inaccessible far side. The photographs were very poor by later standards, but a few features could be clearly identified, of which the most prominent crater was named, fittingly enough, for the Russian rocket pioneer Konstantin Tsiolkovsky.

By early 1961 the Soviets were ready to make the first attempt to reach another planet. The escape velocity, which corresponds to the kinetic energy needed to be imparted to a body in order for it to enter orbit round Earth, is 11 km/s or 40,000 km/hr. To achieve an interplanetary trajectory, a little more oomph is needed – for

A mock-up of the Soviet Venera 1, the first probe to Venus, in the Memorial Museum of Astronautics in Moscow.

Venus, it is 45,600 km/hr. At that point the spacecraft has become an independent body travelling in its own orbit around the Sun. The path by which it coasts to the other planet is referred to as the transfer orbit; in order to reach a planet lying inward towards the Sun, such as Venus, the spacecraft must be slowed relative to Earth; to travel to one further out, like Mars, it must be sped up.

Two Venus probes were launched from Baikonur in February 1961, and marked the beginning of what was to become the Soviets' enduring fascination with the planet. The first, now designated Venera-IVA No. 1, was not publicized at the time, and failed to leave Earth orbit. The second, Venera 1 (from the Russian word for Venus) set out successfully on 12 February 1961, and remained in contact for about a week, by which time it had reached a distance of some 1.9 million km from Earth. At that point, probably owing to a malfunction of its orientation system, radio contact was lost. Venera 1 is thought to have passed within 100,000 km of Venus on 19 May, but no useful data was obtained. Though it ultimately failed, it made history, and will always be remembered as the first spacecraft to attempt to reach Venus.

Targeting Venus, the Americans had the first success. By the summer of 1962, they had readied two spacecraft. The first, Mariner 1, failed minutes after launch on 22 July. Its flight spare, Mariner 2, set out on 27 August.[5] After a three-and-a-half-month voyage across interplanetary space, it reached Venus, 58 million km from Earth, on 14 December 1962. During its flyby of the planet, its radio signal was picked up by a 26-metre tracking antenna of NASA's Deep Space Network (DSN) at the Jet Propulsion Laboratory-operated Deep Space Network research and development station in the Mojave desert, California, and by the Parkes Dish in Australia. It was the latter's first planetary reception and there then began a long collaboration with NASA. For the first time in human history, scientific data was being received from the vicinity of another planet.

The 26-m antenna at the Goldstone 'Venus' station, used to detect the radio signal from Mariner 2 during its 14 December 1962 flyby.

During its flyby, before heading off and out of radio contact into orbit round the Sun, Mariner 2 returned a treasure trove of data. Among its results was an important negative one: unlike Earth, Venus has no global magnetic field. Given Venus's just-discovered slow rotation, this was hardly surprising, as the existence of a magnetic field is thought to require a molten core and a reasonably rapid rotation. But the question it raised was, how does Venus retain its atmosphere in spite of being bombarded by solar wind? Though the spacecraft did not carry any cameras, on the grounds that Venus's clouds were nearly featureless, it was equipped with microwave and far-infrared radiometers, whose measures of microwave radiation (the region of the spectrum beyond the infrared) would show definitively that Sagan's 1960 result had indeed been on the right track.[6]

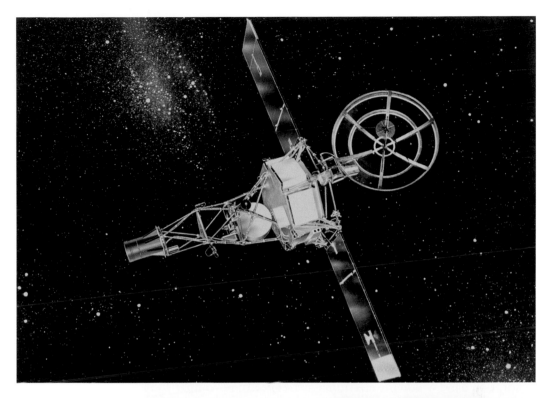

Artist's conception of
Mariner 2 en route to
Venus.

President John F. Kennedy is
shown a model of Mariner
2 by NASA officials, including
William H. Pickering,
centre, director of NASA's Jet
Propulsion Laboratory, at a
meeting in the White House
in 1963. NASA Administrator
James E. Webb is standing
directly behind the model,
and Bob Parks is standing
to Pickering's right.

Before Mariner 2, there were two possibilities. Some astronomers believed that Venus had an electrically charged ionosphere causing high-temperature readings even though the surface was actually cool. In that case, as the spacecraft observed the limb, the instruments would see more of the electron-charged ionosphere and little, if any, of the cooler surface, and the readings would show 'limb brightening' – increasing temperatures as the spacecraft neared the planet's edge. In the other case, which corresponded to the belief of Sagan and most astronomers, Venus had a hot surface but no heavy concentrations of electrons in the atmosphere. The spacecraft would be looking from space at a hot planet, covered by cold, thick clouds. Straight down, the radiometers would register the hot surface through the clouds; at the limb the so-called X band of frequencies in the microwave radio spectrum would be absorbed by sulphuric acid vapour/droplets, producing 'limb darkening', corresponding to decreasing temperatures towards the edge. The instruments looked at Venus at wavelengths of 13.5 and 19 mm. The 13.5 mm band had been chosen because it was the location of a microwave water absorption band within the electromagnetic spectrum; comparisons with the 19 mm band, which was unaffected by water vapour, would allow the detection of water vapour, if there was a significant amount in the atmosphere of the planet. The scans did not reveal the presence of water vapour, but did indicate a definite limb-darkening effect, implying a fairly uniform surface temperature on both the day and night hemispheres of 427°C (800°F). In sum, not only were the high temperatures previously suggested by radio observations quite real, it was clear that they could only result from a greenhouse effect in a massive, mostly carbon dioxide atmosphere, just as Sagan had proposed. (As a jest, Sagan had previously made a $100 bet with astronomer Gérard de Vaucouleurs that the surface pressure on Venus would prove to be 100 times that of Earth, and Mariner 2 won it for him. De Vaucouleurs paid up, while offering as an excuse that his area of expertise was Mars, not Venus.[7])

The first direct measures of the temperature, pressure and density of Venus's atmosphere were obtained by the Venera 4 probe on 18 October 1967. After entering the atmosphere it continued to send data down to an altitude of about 25 km above the surface, when communication was lost.[8] Profiles of the temperature vs height above the surface and of pressure vs height were also obtained during the Mariner 5 flyby mission, whose closest approach to the planet occurred a day after Venera 4's descent through the atmosphere. The next results were obtained from the Venera 5 and 6 landers in January 1969. Though it was claimed at the time that the two landers continued to communicate all the way to the surface, subsequent analysis indicated that Venera 5 went silent at an altitude of 12 km above the surface, Venera 6 at an altitude of 18 km. From their temperature, pressure and density profiles, the surface temperature was estimated as 497 ± 60°C and the pressure as 67 to 135 bar, with 90 bar being the most likely value; the latter corresponds to the pressure at a depth of some 1,000 m below the surface of the Earth's ocean.[9]

Greenhouse Warming on Venus and Earth

On Earth, greenhouse warming has prevented the planet from becoming a permanent ice ball, though according to palaeontologists, it has apparently entered long periods of deep freeze in the past, as during the Cryogenian period of the Neo-proterozoic eon, between 750 million and 635 million years ago (possibly triggered by rapid weathering of rocks associated with the break-up of the supercontinent Rodinia followed by worldwide deposition of the carbonite mineral dolomite which depleted carbon dioxide from Earth's atmosphere). The 'Great Dying' at the end of the Permian period, 252 million years ago, when 75 per cent of Earth's marine life, including trilobites, were completely lost, and many species including rhynchonelliform brachiopods, crinoids, shelled cephalopods such as nautilus and snails suffered severe losses, may also have been due to global warming. The mass extinction seems to have been triggered by an asteroid impact, forming the recently identified 200-kilometre Bedout Crater in northeastern Australia, which set off a series of continuous volcanic eruptions in the area known as the Siberian Traps with a massive output of carbon dioxide leading to an abrupt rise in temperatures combined with a drop in oxygen levels.

The last time carbon dioxide amounts were as high as they are now was more than 3 million years ago, when the temperature was 2–3°C (3.6–5.4°F) higher than during the pre-industrial era, and sea level was 15–25 m higher than today. Recalling the Permian extinction and concerned with the current build-up of carbon dioxide in Earth's atmosphere, palaeobiologist Peter D. Ward sees it as 'a road we humans might again travel on now, seemingly oblivious to the road washout ahead, an accident about to happen one more time'.[10] It is a road that Venus has already travelled.

Venus's massive atmosphere consists almost entirely of carbon dioxide, with the remainder consisting of nitrogen and trace amounts of other constituents, including noble gases, as shown in the table:

Carbon dioxide	96.5%
Nitrogen	3.5%
Sulphur dioxide	variable
Water vapour	variable
Carbon monoxide	variable

The venusian atmosphere traps in heat with efficiency: even though only 2.5 per cent of sunlight falling on the upper cloud tops reaches all the way to the surface, the planet is far warmer than any kitchen oven and half as hot as a blast furnace. The temperature reaches 477°C (890°F) in low-lying areas of the planet, and even on the highest peaks of the planet's mountainous surface remains above 377°C (710°F). Lead, if present on the surface, would melt.

The massive atmosphere of Venus has been invoked as a possible reason for the planet's wrong-way rotation. Though we know nothing certain about early Venus's rotation, it is reasonable to assume it once rotated normally, that is, with its spin axis aligned with the Sun and with a short period of spin. It is also likely that Venus originally had water oceans like those of Earth (again, not certain; some claim clouds on the night side would have kept the surface too warm at night for oceans to form). At some point, the oceans were lost, and the planet developed its massive, sloshing, viscous, almost-pure carbon dioxide atmosphere. Core-mantle friction and solar thermal tides in this atmosphere began to have their inevitable effect on the surface. Simply put, the thermal heating from the Sun holds in position a portion of the atmosphere which becomes extended into the form of an ellipsoid. Beneath these tidal bulges the globe of the planet continues to rotate. If the atmosphere behaved as a frictionless fluid, the long axis of the ellipsoid would remain always oriented directly towards the Sun. Friction at the surface with mountains acts effectively as a brake. On Earth the lunar tides have caused the rotation to slow down from perhaps only a few hours when it formed (the calculations are necessarily highly uncertain for the earliest period) to about 23 hours at the time of the dinosaurs to 24 hours today. The Moon formed about 10 or 20 times closer than it is today and since has slowly retreated out to its present distance.[11] Venus, of course, has no moon, at least at the present time. Nevertheless, at some stage friction between its massive (and lagging) atmosphere and the rotating solid body of the planet would have become so great that, even if it once rotated normally, it would have slowed to the point where it came to be in 1:1 spin-orbit coupling with the Sun – that is, its day and period of revolution would have been equal – only to overshoot and continue to slow down until it ended up in its currently, apparently stable, situation.[12]

As soon as radio astronomers refined the value of the rotation of Venus to 243.01 days, they recognized that this is very close to 243.16 days, the period of a resonance with Earth in which Venus would make, on average, four axial rotations as seen by an Earth observer between successive close approaches of the two planets. Whether these alignments are more than coincidence remains open to debate, but it is possible, if the rotation period of Venus changes over long periods of time (as does Earth's), that Venus's rotation may become 'captured' with Earth for periods.[13]

SEVEN

WHAT ARE THE CLOUDS MADE OF?

Sometimes hath the brightest day a cloud.

SHAKESPEARE, *KING HENRY VI*, PART II

Mariner 2's journey past Venus in 1962 marked humanity's first successful attempt at spacecraft reconnaissance of another planet. However, it was not equipped with a camera; nor were the next visitors to Venus, Mariner 5, which completed a successful flyby in October 1967, and Venera 4–8 probes. It was not until 1974 that the planet's global cloud cover was revealed in detail, when Mariner 10 made a flyby of Venus (minimum distance 5,768 km) during a gravity-assist flyby towards Mercury, and sent back a series of high-resolution television images of Venus in ultraviolet and orange light.

Approaching from the dark side of Venus, the early crescent images did not reveal any detail. After closest approach, on the night side, the day-side images showed amazing contrasts all over the planet, organized on large scales and disorganized on small scales, imaged at increasing spatial resolution as the spacecraft receded from the planet, and eventually showed the morning terminator with reduced contrast. The spacecraft's ultraviolet images showed a plethora of forms at different latitudes. In polar latitudes, the clouds had the appearance of thick stratus, that is, with close to a horizontally uniform base and top. At 45–50° latitude, the clouds looked like those strewn along a cold front in an extra-tropical cyclone on Earth. In lower latitudes, they resembled cumulus clouds, breaking up into a mottle of dark and light blobs. Directly beneath the Sun (the sub-solar region) they gave the impression of strong convective activity. The orange filter images showed far more muted contrasts. Some images in polarized UV light were also obtained.

Additional imagery of the clouds has been obtained by later spacecraft, including NASA's orbiter Galileo during a gravity-assisted flyby of Venus en route to Jupiter, NASA/ESA's Cassini during two flybys en route to Saturn, NASA's MErcury Surface, Space ENvironment, GEochemistry, and Ranging (MESSENGER) en route to Mercury, and most notably the highly productive and enduring orbiters ESA's Venus Express and the Japanese Aerospace Exploration Agency's (JAXA) Akatsuki.

The black-and-white highly stretched UV images from Mariner 10 suggested that the planet is marked by prominent stripes and bands. In unstretched data, however, the patches are grey, and the true appearance is more like a bland gas giant such as Saturn as seen through a telescope. The patchiness in brightness is genuine, and reveals the presence of varying concentrations of ultraviolet absorbers of some kind. Verner E. Suomi, the 'father of satellite meteorology' and a member of the Mariner 10 Venus–Mercury imaging team, referred to them as 'cumulus bituminous' as the identity of the absorbing materials has still not been discovered. Curiously, there do not seem to be any differences in the physical properties of the cloud particles in dark and bright areas inferred from available data, so an outstanding question is why these contrasts occur at all – why aren't the absorbers well mixed in the atmosphere? Also, no consistent differences are noticeable in the measured motions of light and dark features in ultraviolet images. Both light and dark areas reveal the same roughly four-day large-scale circulation of the atmosphere measured by Boyer and Guérin from telescope images from Earth. The upper atmosphere

Verner E. Suomi, known as the 'father of satellite meteorology', was the only atmospheric scientist on the Mariner Venus/Mercury mission (Mariner 10). He was also the principal investigator for the Small Probe Net Flux Radiometer on the three small probes of the Pioneer Venus Multiprobe mission.

is measured to be superrotating at a rate some sixty times faster at cloud tops than the planet itself, a situation that was so hard for astronomers to understand that some of them were unable to convince themselves it could possibly be true. Instead they proposed that the superrotation was only an illusion, and that what they were actually witnessing was the mere pattern speed of some kind of wave phenomenon.[1]

Next off to Venus were the radar-mapping Pioneer Venus orbiter in May 1978 and about a month later the Pioneer Multiprobe mission. The latter consisted of a carrier bus spacecraft, one large probe and three smaller atmospheric probes (named North, Day and Night) that were released from the carrier some two months before arriving at Venus, and descended into the atmosphere on 9 December 1978.

Further data was obtained by the Soviet VeGa 1 and 2, massive spacecraft of Venera design, each weighing 4,920 kg, and representing one of the most ambitious interplanetary missions ever attempted. The initial Venus-only mission was changed to take advantage of Halley's Comet's return to perihelion in 1986 (hence the name, VeGa being a contraction of the Russian words *Venera*, for Venus, and *Gallei*, for Halley). Launched in December 1984, the two spacecraft made close approaches to Venus in June 1985. Two days before closest approach, each flyby spacecraft released a descent lander, similar to previous Veneras, but with different instruments, as well as a balloon with a payload of instruments to study the atmosphere and clouds, while the carrier part continued on to Halley's Comet. The VeGa 1 lander came down successfully in

135

Rusalka Planitia, VeGa 2 in Aphrodite Terra. The balloons with their payloads were released from their compartments on the landers by means of a series of intricate parachute manoeuvres.

In 1970, the radio signals from Venera 7 and Venera 8 entry probes were received by the United States and the Soviet Union independently, and the Doppler showed winds (along line of sight) in excess of 120 ms-1 above 46 km altitude and decreasing rapidly to about 40 (Venera 8) and 20 ms-1 (Venera 7) at 38 km altitude and both varying little for the next 20 km, down to about 18 km and then decreasing to zero at the surface.[2] These were line-of-sight winds measured by the Doppler shift of the telemetry signal, which was close to the east–west direction (zonal) but at some

Model of the VeGa probe bus and landing apparatus. Udvar-Hazy Center, Dulles International Airport, Chantilly, Virginia.

Model of VeGa balloon and payload basket containing instrumentation at Udvar-Hazy Center, Dulles International Airport, Chantilly, Virginia.

small angle, so the measurements reflected a small component of the north–south wind. The four Pioneer Venus small probes were tracked using very long baseline interferometry, allowing both the east–west and north–south components to be measured; the technique had enough accuracy to also obtain the fall speed of the probes. Closer to the surface, the wind direction was highly variable. Venera 9 and 10 landers measured the wind speed at about 1.5 m above the surface to be about 1 m/s, but the direction was not measured. Although this may seem like a very light wind, keep in mind that with the density near the surface about 65 times that on Earth, even this light breeze packs quite a wallop.

The constant level VeGa balloons were floating at an altitude of 53.6 km on the planet's night side, which put them in the most active layer of Venus's three-tiered cloud system. Injected about 1,200 km and two days apart, the two balloons began drifting westward in the zonal wind flow, crossed the day/night terminator, and finally lost communication on the day side after travelling about a third of the way round the planet's circumference. The VeGa 1 balloon's average speed was about 69 m/s (248 km/hr) and VeGa 2's about 66 m/s (238 km/hr). This was consistent with the earlier Venera results, and proved once and for all that the superrotation does indeed extend to the night side deep below the cloud tops. The VeGa balloons also detected a hint of the solar thermal tide component in the winds.

By convention, winds are described in terms of the direction from which, rather than towards which, they blow, and thus westerlies are west to east, easterlies east to west. On a tilted planet like Earth, where the direction of rotation is west to east, westerlies are prograde and easterlies retrograde.[3] Superrotation simply refers to whether the rotation is faster or slower than the underlying solid planet or core. Venus is not the only planet that shows superrotation. Jupiter and Saturn superrotate at the equator but not at higher latitudes. Earth also shows occasional superrotation in the upper atmosphere.

What makes the superrotation on Venus unusual is not the absolute wind speeds, since on Earth the jet streams embedded in the mid-latitude westerlies blow at twice the speed of Venus's winds (at well over 500 km/hr) but the anomalously high speed of its winds relative to the solid body of the planet. On Earth, the atmospheric superrotation of the mid-latitude westerlies is balanced out to within a few per cent by the subrotating tropical easterlies. This means that the net angular momentum of Earth's atmosphere comes out to within a few per cent of that expected for an atmosphere co-rotating with the surface. On Venus, the superrotating equatorial region carries an excess of angular momentum of some tenfold.

Rather surprisingly, the first spacecraft image of Venus in reflected white light, corresponding to the view of visual observers with telescopes from Earth, was obtained only in 2006, by NASA's MESSENGER spacecraft on its way to Mercury. There is no appreciable detail, showing again why Earth-based visual observers found their studies of the planet so uninformative. Clearly, much of what they had seen, although perhaps not all, had been illusory.

We now know a only a little bit more about the Venus clouds. The cloud cover shows different morphologies at different wavelengths and at least at UV there appears to be a recurring pattern. The visible sulphuric acid cloud deck, with whatever absorbers give rise to the dark markings (prominent at ultraviolet and weaker at longer wavelengths) are at heights of 65–70 km above the surface, where the temperatures range from about -43°C (-45°F). Above these cloud tops, there is a pervasive haze of small particles. The contrast features in the cloud cover move nearly east to west (in the same sense as the solid planet's spin at speeds around 100 ms-1 near the equator and often increasing to 120 ms-1 at mid-latitudes). While the cloud motions, taken as proxy for winds, have been reported to vary over time, it is not known yet whether the change is real or due to sampling different phase

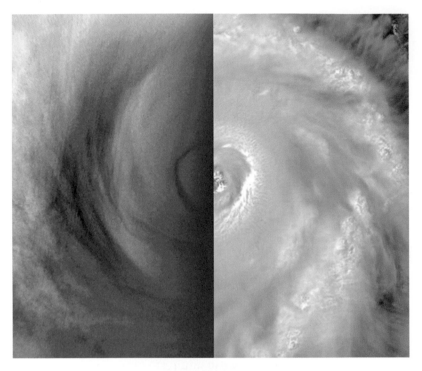

Dynamical and morphological similarities between vortex circulation on Venus (left, from Venus Express data) and a tropical cyclone/hurricane on Earth. Given the challenges in measuring the deep atmospheric circulation of Venus, the morphological similarities provide clues towards understanding the processes involved in the maintenance of Venus's atmospheric superrotation.

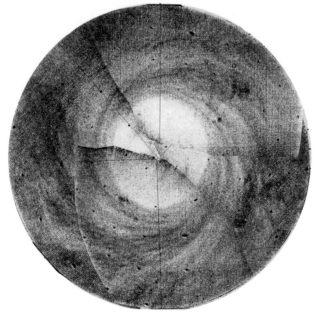

Space–time composite view of the southern hemisphere of Venus constructed from Mariner 10 images showing the vortex organization of the atmosphere situated over the southern pole.

angles or varying cloud altitude, since the concurrent thermal structure measurements are too sparse to reveal the expected change in temperature with latitude at the level of the cloud motions.

At high latitudes are large-scale jets, one in each hemisphere at about 45° latitude, which are also variable; a bright collar, or polar swirl (the reality that underlies the 'polar caps' or 'cusp caps' familiar to visual observers). There appear to be times when the polar jets are absent altogether, and the cloud tops follow the rigid-body rotation. Verner Suomi and Sanjay Limaye first showed that the global cloud cover was organized in two giant vortices situated over each rotation pole of the planet.[4] This organization is fundamental, has been noted in all subsequent observations, and bears a dynamical and morphological resemblance to the tropical cyclones/hurricanes on Earth.[5] High-resolution views show the cloud deck has an almost laminar (eddy-free) flow in the morning sector but becomes patchy and mottled in low latitudes in the afternoon. This structure is consistent with the formation of convection clouds produced by solar heating. The low-latitude mottled clouds appear to have no particular alignment or orientation, but as they migrate polewards they become sheared and stretched from blob-like forms into linear streaky features, which become the dominant forms at mid- to high latitudes (the tilt of the streaks is related to shear stress and the length of time it takes for a cloud patch to drift to mid-latitudes). These include such regularly recurring features as the classical Y-shape first recorded in the Boyer and Camichel era.[6]

The Role of Sulphuric Acid in Venus's Clouds

We know that the main constituent of Venus's atmosphere, by far, is carbon dioxide (96 per cent, followed by nitrogen, 3.5 per cent, carbon monoxide, argon, sulphur dioxide and water vapour, less than 1 per cent). Curiously, the atmosphere does not appear to be

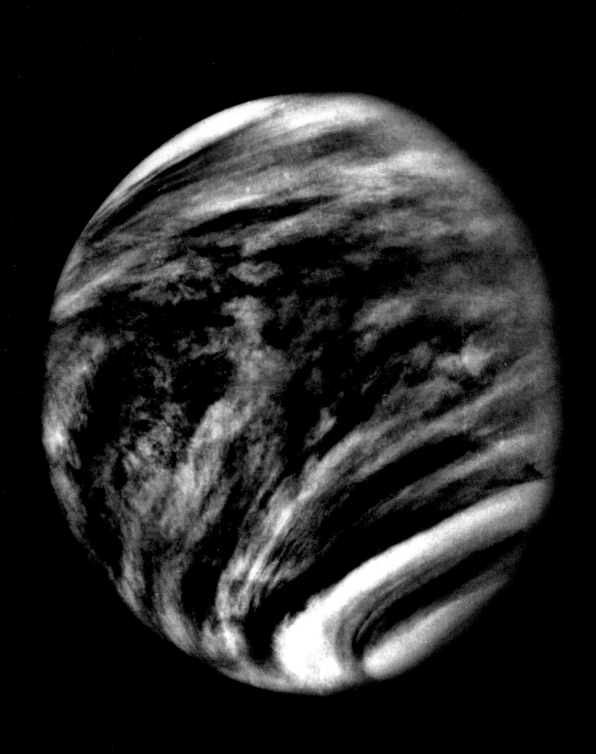

Left: a newly processed version of the Mariner 10 image from 7 February 1974, created using modern image-processing software with archival data. This is a false-colour composite created by combining images taken using orange and ultraviolet spectral filters on the spacecraft's imaging camera. The image on the right is a contrast-enhanced version.

Venus, showing ultraviolet markings in the upper atmosphere, photographed by Mariner 10 en route to Mercury with an ultraviolet filter on 7 February 1974. The whitish belt bordered by a dark collar near the bottom of the planet's image corresponds to Gruithuisen's polar cap and collar.

well mixed in the bulk composition, however, as ~2% variations with altitude in the nitrogen abundance have been detected below 60 km. Carbon dioxide is a colourless gas. The cloud tops, on the other hand, are pale yellow, so clearly they must be formed of something else – and the something else is mainly sulphuric acid droplets (with, as we now know, traces of hydrochloric and hydrofluoric acids and other suspected salts).

On Earth, sulphuric acid, formed by atmospheric oxidation of sulphur dioxide in the presence of water related mostly to human activities and volcanoes, is an important constituent of acid rain. On Venus, sulphur dioxide is present in the atmosphere; however, its abundance above the clouds has been known to vary over decades, which has led to speculation that it may originate in volcanoes still active on the surface of the planet. For an active volcano on Venus to produce an explosive eruption like those of Vesuvius, Krakatoa or Pinatubo on Earth, it would require eruptive force and a fair amount of water, otherwise it would simply ooze lava gently onto the surface, rather as the volcanoes on the bottom of Earth's oceans do. Though one can imagine sulphur gases and sulphur dioxide rising and combining high in the atmosphere to create the sulphuric acid droplets that form the clouds, since the clouds lie at heights of 48–70 km above the surface, the difficulty is getting them to rise to such levels, especially since sulphur dioxide is much heavier than carbon dioxide and nitrogen – the main constituents of Venus's massive atmosphere. At the moment, these matters are still not entirely understood; nor do we even know the nature and the identity of the absorbers mixed within the clouds and cloud particles, though some kind of sulphur iron compounds, including microorganisms, seems plausible.

Whatever may be going on at higher elevations, we know from spacecraft that in lower regions sulphuric acid rain likely occurs, but the temperature of the atmosphere rapidly becomes so hot below the clouds that the droplets evaporate well before they reach the ground. At heights of less than 31 km, which includes 90 per cent of the atmosphere by volume, the atmosphere is clear, though some haze layers have been reported.

Going Over to the Dark Side

Little was known about the night-side cloud cover until 1983, when David Allen and John Crawford discovered details serendipitously while testing a new infrared detector at the Anglo-Australian Telescope at Siding Spring, New South Wales.[7] Their images taken between 1,000 and 2,500 nm showed features analogous to the ultraviolet features seen in reflected sunlight. The features are understood to be due to differences in the opacity of the overlying clouds, so that some of the thermal radiation from the hot lower atmosphere and the surface is able to escape to space at some wavelengths.[8] These atmospheric windows (at 1,100, 1,180, 1,270, 1,740 and 2,300 nm) to the escaping radiation from the lower atmosphere and surface have now been exploited by Galileo, Venus Express and Akatsuki missions to observe the Venus night-side

This global view of Venus is a mosaic of several images taken by the Visible and Infrared Thermal Imaging Spectrometer (VIRTIS) on board ESA's Venus Express on 18 May 2007, at a distance of about 66,000 km from the planet. The images were obtained at 1.7-micrometre (left) and 3.8-micrometre (right) wavelengths. The wavelength used to obtain the left-hemisphere composite (1.7 micrometres) provides a dramatic global view of the night-side clouds in the lower atmosphere (approximately 45 km), while the wavelength used to obtain the right-hemisphere composite (3.8 micrometres) provides a view of the day-side cloud top (approximately 60 km). The day side (right half) low latitude portions are saturated due to the reflected solar radiation at 3.8 micrometres. The hemispheric vortex roughly centred over the south pole is visible at the centre of the image.

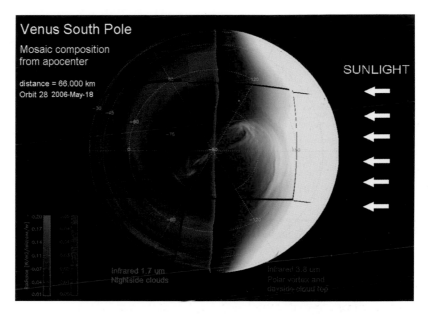

cloud cover. Near 1,100 nm, the radiation from the surface is able to escape through the thick clouds, while at other window wavelengths the escaping radiation comes from altitudes between about 10 and 30 km above the surface. The first image of the surface of Venus at 1,000 nm had previously been obtained from Earth with the 100-centimetre Pic du Midi telescope in 1991.[9] Since then a number of amateur astronomers have been able to achieve the same result using much smaller telescopes.

The surface was also imaged through these windows by the Visible and Thermal Imaging Spectrometer (VIRTIS) and the Venus Monitoring Camera (VMC) on the Venus Express orbiter. These observations led to searches for recent volcanic activity on Venus.[10] However, the results have been suggestive, not conclusive, since the polar eccentric orbit of Venus Express was not suited to covering the planet sufficiently. The Akatsuki data in particular shows a very dynamic cloud cover on the night side. Sadly, both the 1 micron and 2 micron cameras stopped returning data after a few months, while the UV and thermal cameras continue to return data as of

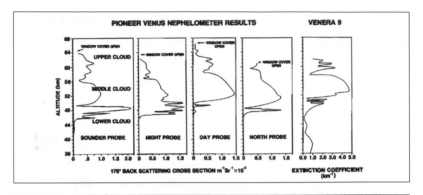

Results of the nephelometers on the four Pioneer Venus probes dropped into the atmosphere in December 1978 and Venera 9 showing the layers of the cloud cover. Although these are instantaneous measurements at different locations and times of day, the general structure is similar with small sub-micron-radius particles in the upper layer, a mixture of the small and about 1-micron-radius particles in the middle layer and lower cloud containing a mixture of 1-micron-radius and larger particles.

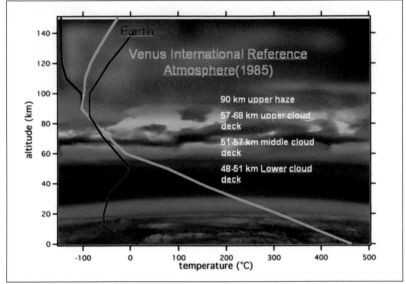

Profile of the atmospheres of Earth and Venus, based on Mariner, Pioneer Multiprobe and VeGa 1 and 2 data.

October 2021. Whereas the cloud cover appears generally very bland in wavelengths longer than violet, the night side of Venus shows incredible detail and puzzling morphologies on a variety of scales.

While the images reveal the horizontal or spatial details of the cloud cover, entry probes descending through the atmosphere to the surface are needed to provide detailed information about the vertical structure of the clouds. However, these investigations have been limited, since the entry probes, after entering the atmosphere at very high speeds, are able to collect data only when they are falling at speeds approaching terminal velocity. Thus the observations generally start at about 64 km, missing the topmost levels of the cloud cover and the overlying haze found at altitudes as high as 90 km, as determined from transit and various Venus Express observations.[11]

At present, the best available data about the vertical structure of the clouds from 64 km to the surface is still that from the Venera landers, the Pioneer Venus small probes and the VeGa 1 and VeGa 2 balloons. An illustration here shows a comparison of the vertical structure at five locations from Pioneer probes and Venera 9 data. Although the probes sampled the clouds at different latitudes and different times of day, the backscatter results generally show three main layers: upper cloud above about 56 km, middle cloud between about 50 and 56 km and lower cloud between about 45 and 50 km. Each layer consists of particles with characteristic sizes, ranging from about 200 nm diameter in the upper layer to about 1,000 nm in the lower layer. However, there are also larger particles, up to 30,000 nm diameter, present. The smaller cloud particles are believed to consist of an aqueous solution of sulphuric acid, with the concentration decreasing with increasing altitude, though with other chemicals also present and perhaps even, as discussed later, microorganisms acting as condensation nuclei.

Open Questions about Venus's Clouds

Among the most important questions remaining unanswered is the identity of the absorbers in the convective appearance clouds, which led Verner Suomi to refer to them as 'cumulus bituminous' (a name that may still turn out to be appropriate if carbon suboxide polymer and graphite turn out to be the absorbers, as proposed by the Japanese planetary scientist Mikio Shimizu).[12] We would also like to know how Venus has managed to keep an atmosphere, unlike Mercury; how the surface pressure has changed and why, unlike

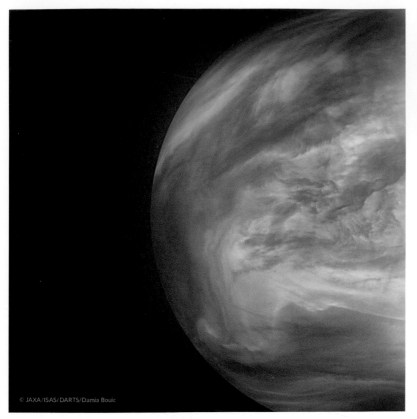

© JAXA/ISAS/DARTS/Damia Bouic

Colour composite of 283 nm, 365 nm and 202 nm reflected-light images taken by Akatsuki orbiter. The bluish polar latitudes (top and bottom of the image) are due to absorption by carbon dioxide in the 2.02 μm image due to depressed cloud tops. The pink and yellow hues depict differences in the amount of sulphur dioxide (absorbs at 283 nm) and some unknown absorbers responsible for the appearance of the 365 nm images.

Earth and Mars, the whole planet is covered with clouds without a break anywhere (and for how long this has been the case).[13] What are the bright and dark patches made of? What are the aerosols below the lower clouds observed by some probes? A challenge for the cloud models is that they must simultaneously transmit solar energy and blanket planetary thermal emission to maintain the apparent high surface temperature. What causes the opacity to change locally on the night side to produce the observed contrasts in night-side images?

Summing up, almost sixty years after Mariner 2, we do know a great deal about Venus's atmosphere. A runaway greenhouse effect explains the high surface temperature. The mechanics of

This image shows the night side of Venus in thermal infrared. It is a false-colour image using data from Akatsuki's IR2 camera in two wavelengths, 1,740 and 2,260 nm. Darker regions denote thicker clouds, but changes in colour can also denote differences in cloud particle size or composition from place to place.

the surprisingly vigorous circulation of the upper atmosphere are now generally well understood, although many of the details have yet to be worked out. The bulk composition is largely known, but the abundances of minor constituents and trace species remain perplexing and poorly defined. The composition of the clouds has also not been fully explained. The nature of the ultraviolet contrasts is unknown, and the identity of the responsible absorbers has proved elusive and remains one of the most intriguing problems concerning the planet.

Perhaps the most important remaining question about Venus is whether it ever had liquid water on its surface, and if so, how and when the water was lost, with Venus instead warming to its current torrid 477°C (750 kelvin) surface with a massive atmosphere almost one hundred times heavier than Earth's? Though some finds indicate that Venus once had liquid oceans, others argue that it never did, and that it has always been hot and bone dry. Unfortunately, the models are very sensitive to initial conditions and assumptions. Future measurements of abundances of noble gases and their isotopes currently planned may finally provide answers, and also determine the compositions of rocks in exposed ancient terrains (tesserae). Then it may be possible to say definitively why the evolutionary paths of Earth and Venus diverged so markedly in the past – one of the great unanswered questions of solar system astronomy.

In having retained its vast oceans, Earth will be prevented from becoming as hot as Venus for some time – though eventually, when the Sun enters the red giant phase of its evolution, our planet's oceans will boil away as well. This will not happen for another 5 billion years (Gyr), so we have plenty of time to revise our calculations. But in the end, we cannot escape the hot conditions already present on Venus, before Venus, and perhaps the Earth too, are swallowed up in the distended atmosphere of their dying star.

EIGHT

THE SURFACE OF VENUS

Talk of mysteries! Think of our life in nature – daily to be shown
matter, to come in contact with it . . . The solid earth! The actual
world! . . . Contact! Contact! Who are we? Where are we?

HENRY DAVID THOREAU,

THE MAINE WOODS, 'KTAADN' (1848)

Venus's surface is like Earth in that it was mapped first at very
high resolutions from the surface before it could be seen in broad
bird's-eye perspective. In Earth's case, the reason for this is that,
until the past sixty years or so, we have been captive to the surface or
the air just above the surface. We have known a planet of rocks, hills,
lakes, oceans, the landscape we see around us, our familiar habitat.
In the case of Venus, though we have seen it as a globe in space, our
gaze has been severely limited by the planet's impenetrable canopy.
The first views of the surface – other than remote ones by radar
whose resolution by 1970 was only about the 10-kilometre scale with
Goldstone and 3-kilometre scale with Arecibo – were provided by
Venera 9 and Venera 10 spacecraft plunging through the lower cloud
layer at about 47 km elevation and thence to the lava plains and rocks
on the surface itself. During this high-resolution view of Venus, the
Soviets were the leaders. Indeed, Venus has been called, not without
justification, the Russian planet.[1]

After several failures, beginning with Venera 1 in 1961, their first real success was Venera 4, which on 18 October 1967 became the first probe to send data about Venus's atmosphere as it descended through the clouds before it gave out, under the harsh conditions of heat and pressure, at about 30 km above the surface. Venera 5 and 6 were of similar but stronger design, but these also failed before they reached the ground. The ultimate goal of reaching the surface and continuing to transmit data after arriving there was finally achieved by Venera 7 on 15 December 1970. It directly confirmed the high surface temperature and pressure (~470°C, 90 Earth atmospheres) inferred from previous probes. Venera 8 followed in July 1972. In addition to other instruments it carried a photometer for the first time. This showed that Venus's clouds reach higher than 62 km altitude and that the atmosphere below this level is relatively clear. Moreover, it showed that at the surface the light level was sufficient for successful imaging, being roughly the same as on an overcast day on Earth and with a visibility of about 1 km. This having been established, the way was clear for the first surface images to be received from the planet, by Venera 9 on 22 October 1975 and Venera 10 three days later. These images were created using a scanning radiometer, much as with the later Viking images on Mars.

Both the Venera 9 and 10 landers survived on the surface for just under an hour before losing battery power. Since the landing sites were located some distance apart, the panoramas they captured showed quite distinct landscapes. Venera 9 landed on a sloping plain, so that the horizon could only be seen in the upper right corner; the landing site was in a debris field consisting of rocks ranging from a few centimetres to tens of centimetres across and some coarse-grained soil. Notably, this represents an increase in resolution by a factor of 100,000 times over the Arecibo radar images. Venera 10, by contrast, showed more flat-topped rocks and finer-grained soil. Significantly, the rocks in both cases, and on Venus generally, are volcanic and basaltic.

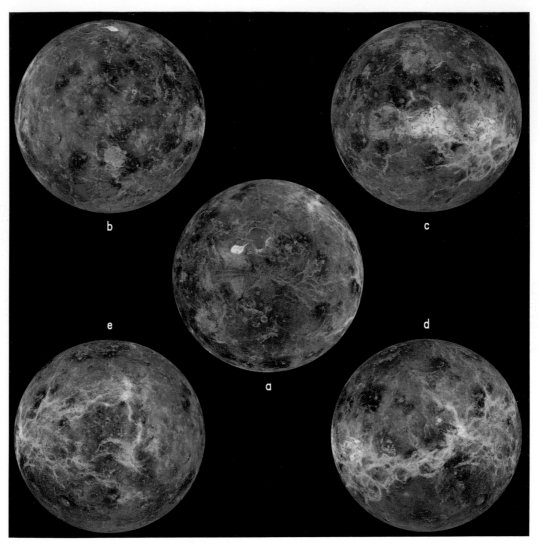

State-of-the-art radar imaging of Venus by Magellan, which was launched aboard Space Shuttle *Atlantis* in May 1989 and began mapping the planet's surface in September 1990. The central image (a) is above Venus's north pole. The other four images are centred on the equator at: (b) 0° longitude, (c) 90° E longitude, (d) 180° E longitude and (e) 270° E longitude. The bright region near the centre in the polar view is Maxwell Montes, the highest mountain range on Venus. Ovda Regio is centred in (c), while Aphrodite Terra, whose terrain is made up of tesserae, is shown in (d). Ovda Regio and Tellus Regio are at the far left, while right of centre is Beta Regio with the two shield volcanoes Ozza Mons and Maat Mons.

Artist's conception of a Venera capsule on the surface of Venus.

Venera 13, which landed on 1 March 1982, and its twin Venera 14, which arrived a few days later, settled east of the highland area known as Phoebe Terra. Venera 13 carried out operations for more than two hours, Venera 14 for about an hour (the planned design life of the landers at the surface was only 32 minutes). They marked a great advance over the previous landers. Each probe used two-colour cameras equipped with quartz camera windows to obtain panoramic images of the surface on opposite sides of the landers. Other instruments included a mechanical drilling arm, which retrieved soil samples, and acoustic microphones able to record atmospheric noise. Venera 9 and 10 landers carried cup anemometers to measure wind speed. The wind speeds at the surface were estimated at between 0.3 and 0.5 m/s, but though sluggish (equivalent to less than 2 km/hr windspeed), the massiveness of the atmosphere gives them considerable heft.[2] NASA had deployed four atmospheric probes from the Pioneer Venus Multiprobe mission of December 1978, one of which (Day, which landed on the day side) continued to transmit from the surface in 1978 for over an hour. Twin Soviet landers VeGa 1 and 2 probes landed one evening in June 1985 near midnight after each deployed a balloon.

Orbiting Radars: Venera 9 and 10 (Bistatic), Pioneer Venus Orbiter, Venera 15 and 16 and Magellan

In addition to their landers, the Venera 9 and 10 as well as Venera 13, Venera 14 (1981) and Venera 15 and Venera 16 (1983) missions included orbiter components, which became the first artificial satellites of Venus. Venera 15 and 16 used radar (32-centimetre wavelength) to map the surface.[3] Venera 15 and 16 orbiters, launched while NASA's Magellan mission was still being developed, carried radar instruments to map the surface as well as an infrared Fourier Spectrometer, which probed the cloud-covered atmosphere. Though only a small fraction of the planet was surveyed and the resolution achieved was no better than the best

Venera 10 panorama of the surface. Both Venera 9 and Venera 10 landed on sloping terrain. The panorama was obtained by scanning the landscape in the vertical plane and rotating it around its axis to obtain a 180° view. Thus the objects in the centre of the frame are seen at highest resolution and the horizon is seen towards the left and right edges.

Undistorted views showing what would be visible to someone standing at the Venera 9 site.

Probe	Venera 7	Venera 8	Venera 9	Venera 10	Venera 11	Venera 12	Venera 13	Venera 14 Lander	VeGa 1 Lander	VeGa 2 Lander	Pioneer Venus Large Probe	Pioneer Venus Large Probe	Pioneer Venus Day Probe	Pioneer Venus North Prober	Pioneer Venus Night Probe
Longitude	351.0	335.3	292.6	291.5	299.0	294.0	303.0	310.0	175.9	177.7	4.4	304	317	4.8	56.7
Latitude	-5.0	-10.7	31.0	15.4	-14.0	-7.0	-7.5	-13.3	8.1	-7.1	304.0	4.4	-31.2	59.3	-28.7

Surface Locations of Venus Entry Probe and Landers

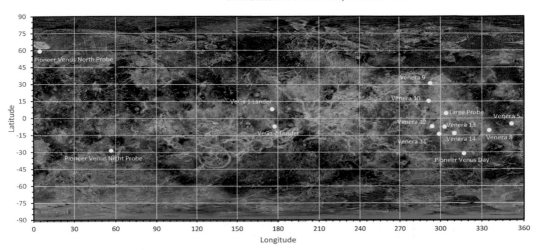

Locations of Venera 8–14 landing sites superimposed on combined Magellan/ Pioneer/Venera topography data.

resolution attained from Earth radars at the time (showing regional differences in surface reflectivity and topography on a scale of ~100 km), it was a significant step forward. A new era had begun: the spacecraft radar mapping of Venus.

Views on intermediate scales, between the very-high-resolution views of the Venera landers and the rough surveys by Earth-based radar and the Venera 9 and 10 orbiters, were contributed by the first American spacecraft to add to the exploration of Venus since the flyby Mariner era. This was the U.S. Pioneer Venus orbiter, which deployed much more sophisticated instrumentation than the Venera orbiters to acquire surface reflectivity and altimetry data during the five years of its mission after it entered Venus's orbit on 4 December 1978. From Pioneer data, a map of the surface of Venus at about 75 km scale was produced, showing global relief but not yet in sufficient detail to reveal the true characteristics of the surface features. It did, however, contribute one important negative finding: there was no evidence of plate tectonics.

Venera 15 and 16 Orbiters

Following the successful Pioneer Venus mission, NASA issued a call for instrument proposals for an ambitious Venus Orbiting Imaging Radar (VOIR) mission. However, facing budgetary constraints at the time, the mission went unfunded. Instead, the Soviets again took the initiative, and in June 1983 launched two high-resolution radar-imaging orbiters, Venera 15 and 16. They were inserted into eccentric polar orbits one day apart in October 1983, with the radar mapping being carried out during closest approaches on successive orbits. They achieved a much higher spatial resolution than had been obtained up to that time (about 1–2 km), together mapping 92.2 million square km of the northern hemisphere over a period of eight months (one sidereal rotation).

Magellan Orbiter

After the brilliant successes of Venera 15 and 16, NASA revived plans for a radar-only mission to Venus following scrapping of the original Venus Orbiting Imaging Radar (VOIR) mission. Thus the Venus Radar Mapper (later to be renamed Magellan) was born. In order to keep down costs (and mollify Congress), a spare antenna from the highly successful Voyager missions to the outer solar system was utilized. Also in order to save money, spare components were taken off the shelf from other missions. Even so, progress was slow because of cost inflation caused by the Space Shuttle programme, which had required that all future planetary missions be launched not on single-use rockets but on Space Shuttles. Unfortunately, the Space Shuttle launch schedule began to lag behind, a situation compounded by the tragic loss of the shuttle *Challenger* in January 1986. When the shuttles began flying again, the Galileo mission to Jupiter received higher priority, and pushed the Venus radar orbiter mission back to 4 May 1989. On that date,

Venera 15–16 reflectivity and altitude image of one quadrant of the northern hemisphere containing Ishtar Terra and Maxwell Montes.

Magellan finally got underway, and followed an energy-efficient but roundabout trajectory round the Sun, finally arriving at Venus fifteen months later, on 10 August 1990. The wait had been long but well worth it. Between then and the end of its mission in October 1994, when it entered and burned up in the atmosphere of Venus, Magellan completed an astonishingly detailed mapping of the surface below the clouds. In all, it completed six orbiting cycles obtaining reflectivity maps at a resolution of about 100–250 m per pixel – a resolution far superior to anything available before. At this resolution, Venus emerged as a strange and unfamiliar world unlike any of the other rocky planets of the solar system. Most emphatically its surface was unlike Earth, so long supposed to be its twin.

It is important to emphasize that though radar images resemble photographs, they differ in some respects. What we see in the radar images is not what the eye sees, and though in both cases reflected electromagnetic radiation is what makes the features visible, the radiation is in different wavelength ranges. At radar wavelengths (a few centimetres to tens of centimetres), the reflected signal depends on the angle of illumination as well as the roughness of the surface on scales comparable to the wavelength. Thus a rough surface appears bright, and a smooth surface dark. When rendered as an image, the appearance will depend both on the roughness and on the incidence angle, so that different systems produce somewhat different results. Venera 15 and 16 maps use an average incidence angle of 10° and the wavelength was 8 cm. Magellan, on the other hand, deployed a side-looking radar, so that the radar signal could be varied with respect to the vertical and the surface could be illuminated from the left or the right. The precisely controlled orbit of Magellan and the long duration compared to the rotation period of Venus allowed the same areas to be covered multiple times at slightly different angles, making stereo images possible and resulting in improved altitude resolution of the topography in the overlapping areas.

Despite the qualitative differences in their imagery, both Venera 15 and 16 and Magellan established clearly that Venus's surface is unlike that on any other planet. Familiar features such as craters, volcanic as well as impact, are present, but there are also many features unique to Venus. Venus has a style of its own. The interior of Venus is believed to consist of a molten iron core approximately 3,200 km in radius. Above this core is a mantle of hot rock slowly churning because of the planet's interior heat, and overlain in turn by a thin crust which the churning mantle catches and drags about. Thus the crust bulges and moves as the mantle shifts, and is piled here and there into plateaus and mountains or moulded by upwelling plumes into volcanoes and rifts.

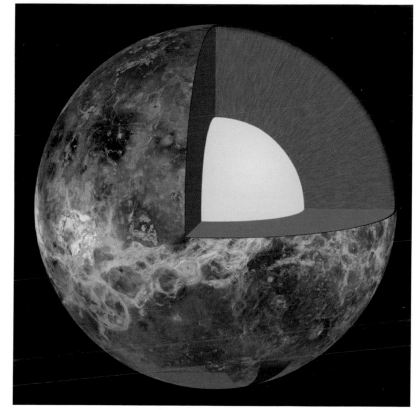

A cutaway view of Venus. Since seismic data is not available for Venus, this model is theoretical, based on similarities between the Earth and Venus such as their densities. Presumably Venus, like Earth, has an at least partially molten nickel–iron core at the centre, a silicate mantle that though solid behaves as a viscous fluid over geologic time (and may partially melt to produce mantle plumes), and the crust. On Venus, there is no evidence of plate tectonics, possibly because the crust is too strong to produce subduction zones without the lubricating effect of liquid water.

Venus's Surface: A Broad Overview

The Venera orbiters and Magellan showed that a large fraction of Venus's surface is covered with lava flows. The lava flows are from volcanoes, and, besides covering the surface, form channels and valley complexes. Hundreds of such formations associated with many of the nearly 2,000 volcanoes found over the planet have been identified from Magellen data. Channels and lava tubes like those found on Earth are common, but those on Venus are on a much greater scale, probably because the basaltic, low-viscosity lavas found in the lava fields on Venus are very fluid and take a long time to cool under the high-temperature conditions. Some channels are simple, and resemble the rills on the Moon, while others, called *canali*, consist of long distinct channels maintaining their width through their entire length; the longest of these, Baltis Vallis, runs for a full sixth of the planet's circumference. Still other channels are complex, and form anastomosed webs similar to those found on Mars. Still others are sinuous or meandering and end in deltas and terminal lakes.

In all, some 80 per cent of the surface consists of smooth, volcanic plains. Both volcanic calderas and impact craters are evident, though in contrast to what we see on impact-battered bodies like the Moon, Mercury and Mars, Venus's surface has relatively few impact craters.

There are significant differences in the forms of venusian impact craters and those on other solar system bodies. The most important is that the impact craters on Venus are less eroded, and smaller ones do not exist at all. On Earth, impact craters have been largely removed by wind and water erosion; on the Moon and Mercury they have been significantly degraded by subsequent impacts. On Venus, though there are hundreds of impact craters up to 280 km in diameter, randomly distributed across the surface, there are none smaller than 3 km in diameter.[4] This is because smaller projectiles entering Venus's dense atmosphere are simply broken up and burned completely before they can reach the ground. Not surprisingly, the ejecta blankets of many impact craters appear markedly stunted; there are certainly none of the great ray systems found in association with fresh impact craters on the Moon which sometimes splay out across thousands of kilometres. On Venus the lobes of ejecta blankets are gentle and round, making them look like splash patterns – which is exactly what they are. Debris ejected from impacts behaves like a liquid on passing through the dense venusian air. The venusian craters have not been subject to erosion by liquid water or ice, and though wind erosion must take place, it does not seem to have been very efficient. But not only do

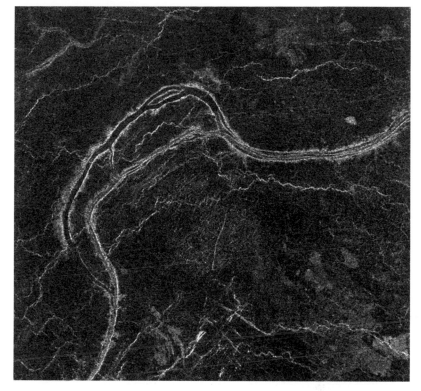

At the left of this image of an area of Sedna Planitia, lava originating from Ammavaru caldera 300 km away (and outside the frame) overflows at the ridge at the left-centre, and pools further to the right. Note the anastamosing, 2-km-wide channel.

the craters appear uneroded. They really do appear to be relatively young. The small number and relatively even distribution of impact craters superposed on the surface seen in Magellan images has been interpreted to mean that the average surface age is between 300 million and 700 million years old.[5] Global geological mapping showed that the first ~80–90 per cent of the geological record was lost through resurfacing, and the subsequent geological history is characterized by decreasing intensity of tectonism and volcanism.[6] Any craters older than around 800 Myr have presumably been obliterated by what has been referred to as a 'cataclysmic global resurfacing event', which has buried the entire older surface under lava flows.[7] Admittedly, some investigators doubt that a discrete resurfacing event occurred and instead believe that the volcanism

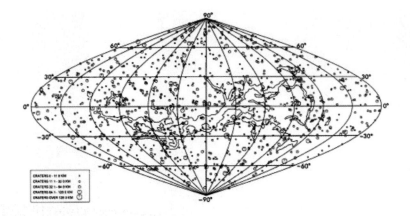

Distribution of impact craters on Venus. From Schaber et al., 'Geology and Distribution of Impact Craters on Venus; What Are They Telling Us?', *Journal of Geophysical Research*, XCVII (1992).

has been ongoing. As with many questions regarding Venus, this one remains open at the moment.

Of the 80 per cent of Venus's surface consisting of volcanic plains, 70 per cent contains low sinuous ridges similar to the wrinkle ridges found on the maria of the Moon; the rest (known as lobate plains) are defined by their numerous distinctive lava flows. The remainder of Venus's surface consists of highlands, first noted in Earth-based radar maps of the 1960s and identified then, incorrectly, as 'continental' areas. There are two main highland areas. The first is Ishtar Terra (named for the Babylonian Venus), at around 60° N latitude. The other is Aphrodite Terra, close to the equator but mostly projecting into the southern hemisphere.

A word about nomenclature. The convention approved by the IAU for features on Venus is to give female names for everything. At the time this convention was adopted (late 1970s), it probably seemed like a courteous gesture on the part of the men who still dominated planetary studies, though in retrospect it seems rather chauvinistic. But for better or worse, it cannot be changed now. The calderas are named for famous women in history; planitiae for mythological heroines; canyons for goddesses of the hunt or moon goddesses; montes for goddesses of any type and so on. But there are three exceptions. Two highland areas on Aphrodite Terra that

The 274-km diameter impact crater Margaret Meade, located north of Aphrodite Terra and east of Eistla Regio. This is a multi-ring crater, with the innermost, concentric scarp being interpreted as the rim of the original crater cavity; the presence of hummocky, radar-bright crater ejecta crossing the radar-dark floor terrace and adjacent outer rim scarp suggests the floor terrace is probably a giant rotated block concentric to, but outside of, the original cavity. Magellan image from 1990.

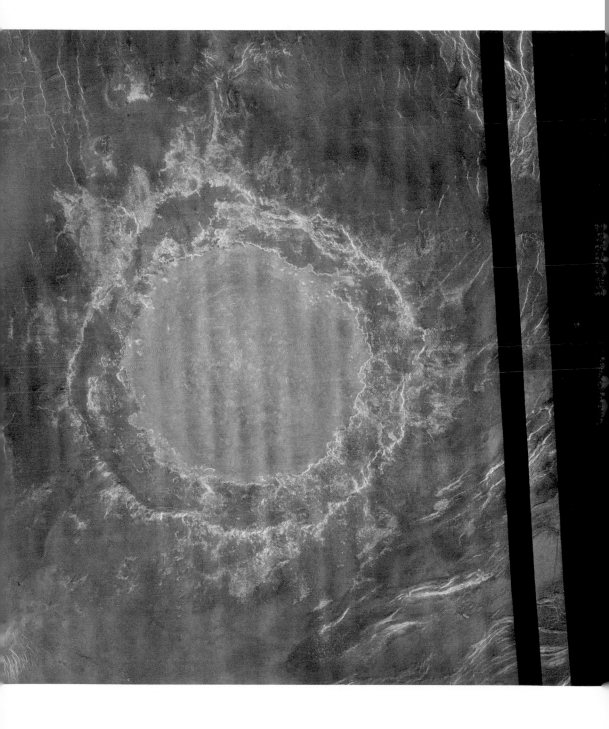

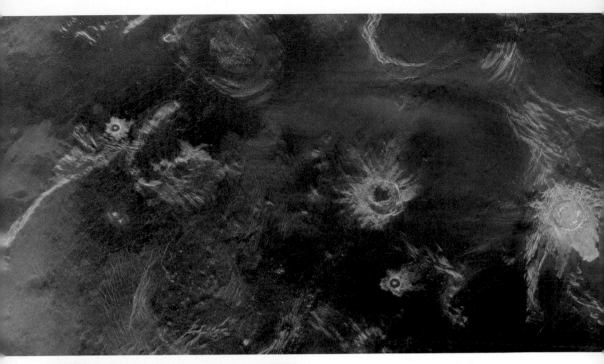

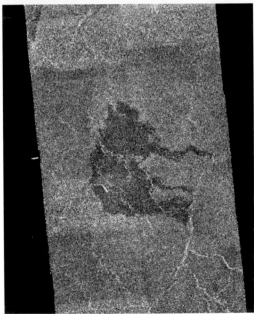

Four impact craters, showing a variety of structures. At left is 29-km diameter Kollwitz, with a prominent central peak and an ejecta blanket barely extending one diameter away. At lower centre is 33-km diameter Callirhoe, also with a prominent central peak and featuring an array of braided channels to the east. North of Callirhoe lies Maria Celeste (97 km), with an interior ring of mountain massifs instead of a central peak, which also shows a braided runout of ejecta, with most of the channels splaying northward from the crater. At far right is Greenway (92 km), with a bright instead of dark floor, which on radar images means a rough instead of a smooth floor. Bright (rough) material has flowed out of the crater forming the fan-shaped lobes towards the southeast, possibly as a result of volcanic activity that began at the site of the older crater.

This Magellan image, covering an area of 75 × 45 km, shows volcanic lava flows overlying brighter and presumably older flows on the surrounding plains. Some of the dark flows terminate on the narrow, bright sinuous feature to the right of the image and it is assumed that they originate from eruptions at fissures located along this bright feature. The horizontal banding is a processing artefact in this photograph made from the first orbit after the start of systematic mapping.

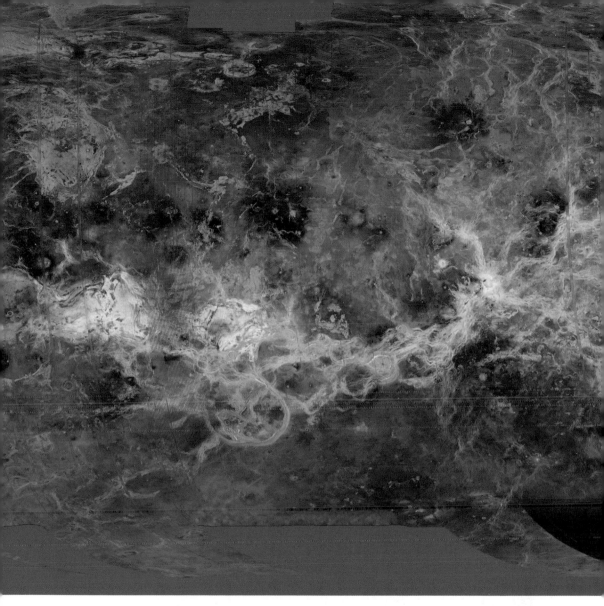

The eastern half of the planet, including Aphrodite Terra, is displayed in this simple cylindrical map of the surface of Venus. The left edge of the image is at 52.5° east longitude, the right edge at 240° east longitude. The top and bottom of the image are at 90° north latitude and 90° south latitude, respectively. Magellan synthetic aperture radar mosaics from the first cycle of Magellan mapping are mapped onto a rectangular latitude–longitude grid to create this image. Data gaps are filled with Pioneer Venus orbiter altimetric data, or a constant mid-range value. Simulated colour is used to enhance small-scale structure. The simulated hues are based on colour images of the surface recorded by the Soviet Venera 13 and 14 landers.

were discovered in the Earth-radar era are still called Alpha Regio and Beta Regio, while the mountain range on Ishtar Terra, also discovered before the IAU convention was adopted, was named Maxwell Montes, after the Scottish physicist James Clerk Maxwell, founder of the theory of electromagnetism on which radio and radar are based. Maxwell is thus – and will doubtless remain – the only male honoured on Venus.

Another IAU convention concerns the prime meridian, relative to which venusian longitudes are determined. The choice is, of course, completely arbitrary. Originally, in the Earth-radar era, this was taken to pass through the oval feature now known as Eve, south of Alpha Regio, but it has now been moved 60 km east. It is now defined as passing through the central peak of the 23.6-kilometre-wide impact crater Ariadne.

In orienting oneself to Venus's map, it is useful to begin with the northern hemisphere highland Ishtar Terra, which was first revealed in detail by the Venera 15 and 16 orbiters. It covers an area about equal to that of Australia or the continental USA. Its western part consists of a high volcanic plateau, Lakshmi Planum (named for the Indian goddess of wealth), which measures 2,500 km across and 3 km high. As there are no seas on Venus, there is no 'sea level', as such; instead the reference level is defined as the mean

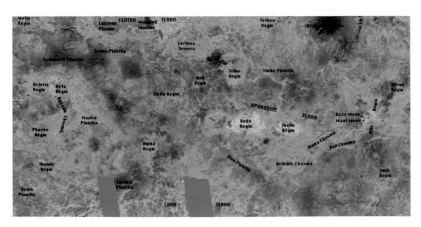

A topographic map of Venus based on Magellan altimetry data, showing some of the main features. The most elevated regions are yellow, intermediate are green and the lowest-lying are blue.

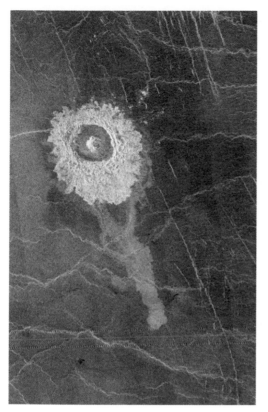

Small Ariadne crater. By IAU convention, the central peak of this crater designates the prime meridian of Venus.

radius, 6,051.2 km. Lakshmi Planum is capped by two volcanic calderas, Colette and Sacajawea, and bordered by a series of mountain ranges: the Akna Montes to the west, the Freyja Montes to the north and the Maxwell Montes to the east. All show clear evidence of faulting and folding and demonstrate that the crust has been subject to intense deformation, like that of Earth. The Maxwell Montes are the highest on Venus, with some of the peaks reaching as high as 9 km above the level of Lakshmi Planum, or 12 km above the venusian reference level, making them higher than the Himalayas of Earth. On the flanks of the eastern Maxwell Montes is an interesting structure, Cleopatra, originally thought to be a volcanic caldera but actually a 105-kilometre-wide multi-ring impact basin.

Alpha Regio, one of the first discrete features discovered using Earth-based radar, is another interesting area. As first revealed by the Venera 15 and 16 radars, it is an example of what is known as a tessera – a type of terrain that is highly deformed and where the deformation strikes in multiple directions and is closely spaced. (The Russian geologist Alexander Basilevsky thought this type of feature resembled a parquet or tiled surface, hence the name.[8]) Tessera terrain appears to be an older section of crust deformed numerous times since its formation, and is often found in association with other peculiar features known as coronae, 'pancake domes' and 'ticks'. Coronae are large ring-like structures, of which about thirty were found in the area round Ishtar Terra by Venera 15 and 16. Each corona consists of an elevated and extensively fractured region surrounded by a ridge, and may indicate the location of hot plumes in the mantle where partially melted material has welled up and caused extension and fracturing of the overlying crust. Alpha is like most such units, in that the surrounding volcanic plains appear to have flowed around its margins, meaning that these plains are younger than Alpha. The 'pancake domes' (officially, farra) are flat-topped structures that appear to have been formed above a plume oriented very close to vertical near the surface, and to have been produced by a more viscous lava than the ordinary basalt that oozed out onto the very flat surface and pushed out against Venus's high atmospheric pressure. On Earth, thicker flows are usually associated with lavas richer in silica, such as dacite or rhyolite, though whether this is the case on Venus is not known; another possibility is that they formed by basalt erupting at unusually slow rates. In any case, the mechanism of their formation, whatever it is, must be peculiar to the local geology. 'Ticks' are domes with flat, concave summits and radiating ridges and valleys on the sides produced by landslides, and sometimes associated with piles of debris; the fact that some of them appear as domical rises with numerous legs explains the name.

Tessera Terrains

Along with Ishtar Terra, the other important highland is Aphrodite, which straddles a third of the circumference of the planet and so, though not quite as lofty as Ishtar, is far more extensive. It consists of a western part, including the two elevated plateaus Ovda Regio and Thetis Regio; a central part cut by a series of ridges and deep troughs; and an eastern part, including Beta Regio, also discovered early on using Earth-based radar.

Diana Chasma in central Aphrodite is impressive by any standard, measuring 280 km across and sinking 2 km below the mean radius, or 4 km below the neighbouring ridges. It marks the lowest point on the entire surface of Venus. Dalia Chasma, nearby, is presumably part of the same system. Beta contains two large shield volcanoes, Ozza and Maat Mons, which lie on a fault line running in a roughly north–south direction and recall shield volcanoes like Mauna Loa and Mauna Kea in Hawaii, though those on Venus are much more massive. Maat Mons is the second-highest mountain after Maxwell Montes, as well as Venus's highest shield volcano. It rises 8 km above the mean planetary radius and nearly 5 km above the surrounding plains, its summit caldera measures 28 × 31 km and it contains at least five smaller collapse craters, of which the largest measures some 10 km across.

In addition to extensive lava flows that cover much of the surface, Venus has a large number of identifiable volcanoes – in fact, more than any other planet in the solar system. Again, there are some contrasts with volcanoes on Earth. On Earth, 'shield' volcanoes, structures with gently sloping flanks, form in places where magma is ejected over hot spots, giving rise to relatively fluid lava from which gases easily escape and build up from repeated eruption of basalt through a single vent. Examples are the Hawaiian islands, which are the tops of giant undersea shield volcanoes. These structures differ from so-called composite or stratovolcanoes, such as Mount Rainier and Mount St Helens, formed by tectonic plate movements in which the oceanic crust of one plate slides

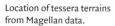

Location of tessera terrains from Magellan data.

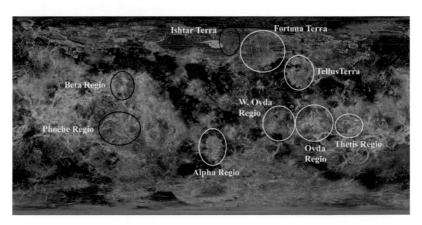

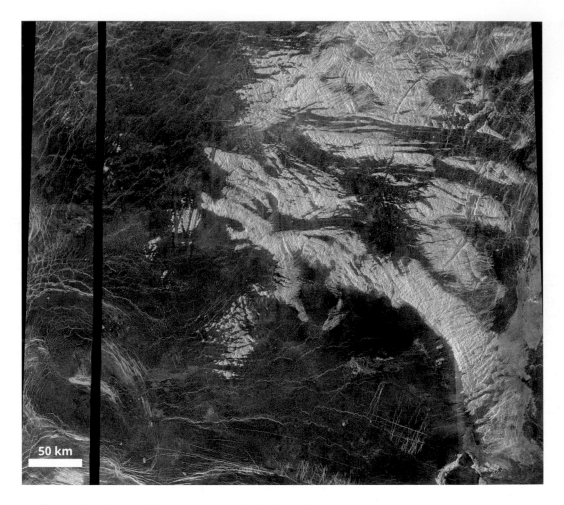

50 km

under another in a subduction zone; the inflow of sea water produces a more viscous lava from which the gas has greater difficulty escaping and builds up to very high pressure, ending in an explosion, but since there is neither sea water nor plate tectonics on Venus, stratovolcanoes do not exist at all.

Even the shield volcanoes on Venus are morphologically distinct from those on Earth. They become much larger than those of Earth, and can cover an area of hundreds of square kilometres. Despite their vast areal extent, their slopes are strikingly gentle. The gentle

An area of Venus located southeast of Fortuna Tessera. The bright, rough terrain is tessera; the dark material between it is smoother lava plains. The plains are cut by polygonal cracks and folded by wrinkle ridges. Plains lavas have flowed into linear cracks in the tessera, indicating that the plains lavas came after the tessera.

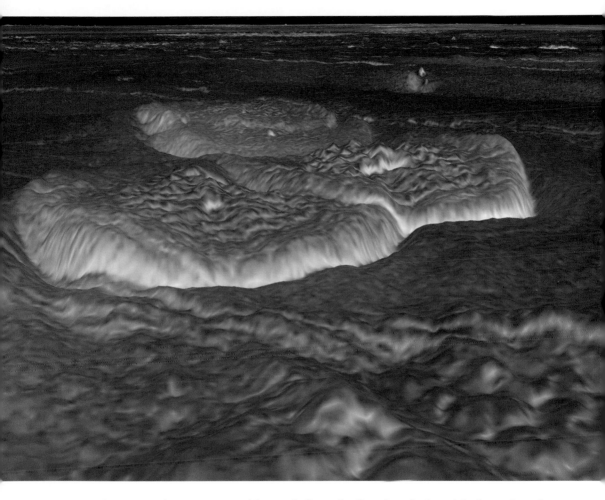

A simulated perspective view of a pancake volcano, created by combining topography data and reflectivity data. Such domes are about 1 km high and tens of kilometres across.

slopes presumably result from the fact that the basaltic, low-viscosity lavas are highly fluid and slow to cool under the high-temperature conditions. Thus shield volcanoes there do not have the structural strength needed to support them vertically. The average height is on the order of only 1.5 km. Nevertheless, they become so massive that they can actually bend the lithosphere (the brittle upper portion of the mantle and crust) downwards, producing flexural moats and/or ring fractures and fracturing magma chambers far beneath the surface. According to one study, there are at least 167 shield

171

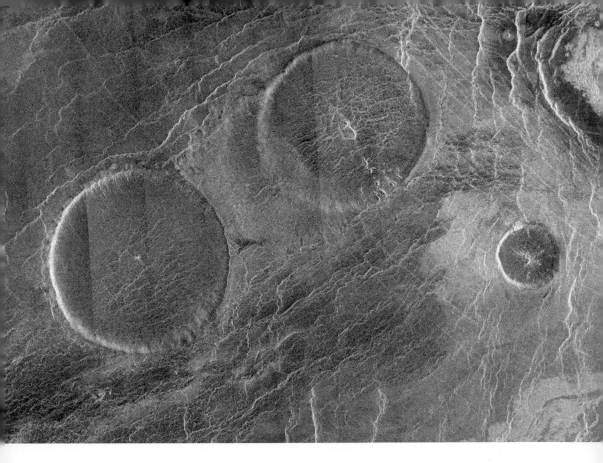

volcanoes more than 100 km across, compared to only a single volcanic complex of similar size on Earth – that in Hawaii.[9] As for the total number, based on Magellan data there are anywhere from 2,000 to 70,000, depending on the criteria used. In addition to shield volcanoes, there are numerous more exotic forms without terrestrial analogues: coronae, characterized by circular or elliptical patterns of concentric fractures; arachnoid volcanoes; pancakes and others.

A Magellan full resolution mosaic, centred at 12.3° north latitude, 8.3° east longitude, shows an area 160 × 250 km in the Eistla region of Venus. The prominent circular features are volcanic 'pancake' domes, 65 km in diameter with broad, flat tops less than 1 km in height. These structures represent a unique category of volcanic extrusions on Venus formed from viscous lava. The cracks and pits commonly found in them result from cooling and withdrawal of lava.

Is Venus Still Volcanically Active?

The ubiquitous lava flows covering Venus's surface, and the countless volcanoes and other volcanic structures (over 1,000 in all), show that volcanism has changed the surface of Venus over the last few hundred Myr. The planet's internal heat has to escape somehow,

Another exotic type of feature on Venus: an arachnoid volcano, so called because features of this type, of which some 250 are known, are made up of concentric ovals surrounded by radial fractures resembling the threads of a spider's web.

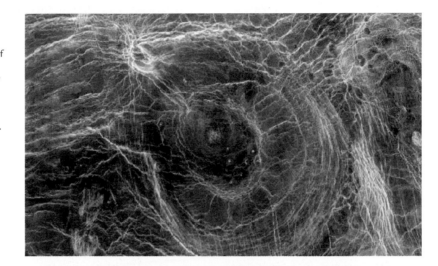

Maat Mons volcano is displayed in this computer generated three-dimensional perspective of the surface of Venus looking south using Magellan synthetic aperture radar and altimetry data. Maat Mons, named for an Egyptian goddess of truth and justice, is located at approximately 0.9° north latitude, 194.5° east longitude. Its peak reaches 8 km above the mean surface. The vertical scale in this perspective has been exaggerated ten times. Rays cast in a computer intersect the surface to create a three-dimensional perspective view. Simulated colour and a digital elevation map developed by the u.s. Geological Survey are used to enhance small-scale structure. The simulated hues are based on colour images recorded by the Soviet Venera 13 and 14 spacecraft.

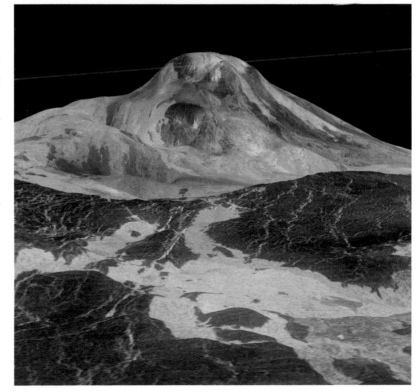

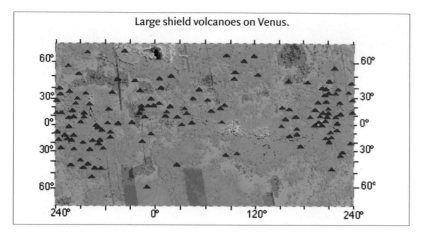

Large shield volcanoes on Venus.

Distribution of large shield volcanoes on Venus.

and has clearly done so through volcanic eruptions and cataclysmic lava flows during much of the planet's ancient past. Whether volcanism is still active has been a question asked often.

Near-infrared monitoring during the Venus night by the ESA's Venus Express (whose mission extended from its orbital insertion in April 2006 until contact was lost in January 2015) showed that the emissions from a few locations were different from those in surrounding terrains. Investigators surmised that this radiation was likely coming from relatively fresh lava flows (less than 2.5 Myr old) that had not yet experienced significant surface weathering.

A provocative suggestion based on variation of sulphur dioxide above clouds over time was presented by Larry Esposito in 1984: episodic injections from erupting volcanoes. This was reinforced in 2012 when investigators reviewing measurements of sulphur dioxide from Venus Express obtained in 2006–7 discovered a sharp rise in sulphur dioxide content had occurred above the clouds, followed by a gradual falling off over the next five years. Since sulphur dioxide is an important component of volcanic outgassing, this could have been due to episodes of volcanic activity. However, sulphur dioxide content in the lower atmosphere remained unchanged, which suggested that the increase above the

An artist's impression of the Venus Express spacecraft orbiting Venus.

clouds could have been merely through the slow overturning of atmospheric circulation.

So far the most suggestive findings were obtained using the Venus Express Venus Monitoring Camera (VMC) with a near-infrared 1,000 nm filter. During several observing sessions in 2008–9, four transient hot spots were detected along the rift zone Ganis Chasma.[10] Ganis Chasma is located in the Ganiki Planitia Quadrangle of Venus, which is between the two volcanic regions Atlanta Planitia to the north – which formed as the result of mantle upwelling and downwelling – and Alta Regio to the south, a major volcanic rise formed as the result of mantle upwelling.[11] It is an extension feature shaped like an arc around the corona Sapas Mons (many of the coronae on Venus are believed to be dormant but not dead volcanoes). Given the lack of associated wrinkle ridges

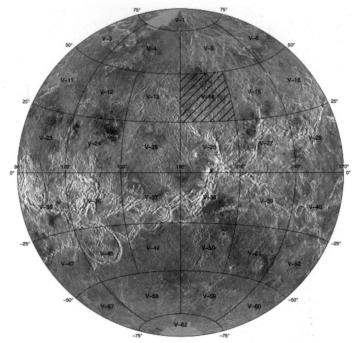

2011

U.S. Department of the Interior
U.S. Geological Survey

Hatched area v-14 indicates the location of Ganiki Planitia, where transient bright spots observed in Ganis Chasma may indicate active volcanic activity on Venus.

it appears younger than other chasmata on Venus. If the claim of infrared flashes at Ganis Chasma can be substantiated, these observations would provide the first direct evidence of current volcanic activity anywhere on Venus.

Recently, modelling of mantle convection and interaction with lithosphere has led to the suggestion that as many as 37 coronae (including Artemis) seen in the Magellan radar data are tectonically active, raising the likelihood that Venus's volcanoes are active presently. Investigating the question has been the Japanese orbiter Akatsuki ('Dawn'). Launched on a what was to be about a six-month journey to Venus in May 2010, it was equipped with a number of instruments designed to confirm the existence of active volcanism, but failed in its initial attempt to enter Venus's orbit in December 2010. Since then it has been a real survivor. It spent five years

Evening hemisphere of Venus is shown in this synthesized false-colour view of Venus on 7 December 2015 from the Akatsuki orbiter. The atmosphere rotates from the limb (right edge) to the evening terminator (left edge) at speeds reaching ~100 ms^{-1} in low and mid-latitudes. This composite-colour image was created with 283 nm filter image (blue); 365 nm filter image (red); and a mixture of the two shown in green. The yellowish areas suggest lesser amounts of the unknown ultraviolet absorbers in the equatorial regions.

Morning hemisphere of Venus is shown in this synthesized false-colour view of Venus on 9 November 2016 from the Akatsuki orbiter. The composite colour image was created with 283 nm filter image shown in blue, 365 nm filter image shown in red, and a mixture of the two shown in green. The amount of SO_2 above the cloud tops is relatively low in the bluish areas in this image.

Full disc view of Venus from Akatsuki orbiter on 6 May 2016 is shown in this synthesized false-colour image. This colour composite was created with 283 nm filter image shown in blue, 365 nm filter image shown in red, and a mixture of the two shown in green. The yellowish areas suggest lower amounts of the unknown ultraviolet absorbers in the equatorial regions.

orbiting the Sun before engineers managed to rescue it, and finally, in December 2015, its long-dormant attitude control thrusters were successfully fired, placing it at long last in an elliptical orbit round the planet.

At the time of writing (June 2021) Akatsuki continues its mission, including the search for lightning associated with atmospheric convection or volcanism. In this investigation it is using a fast-sampling optical sensor designed to measure the light curve of lightning flashes and appears to have detected one probable event.

Artist's concept of lightning on Venus. The magnetometer on board the ESA's Venus Express orbiter detected wave signals that show evidence of lightning in the atmosphere.

A Flotilla of Planned Spacecraft

Though Akatsuki obtained some data during its approach to Venus in 2010, it began regular monitoring in early 2016. It has continued to make observations from Venus. Since 2014, when communications with the ESA's Venus Express were lost, Akatsuki has had the planet almost to itself. The only other spacecraft reconnaissance of Venus has occurred incidentally during gravity-assist flybys of spacecraft with other priorities: NASA's Parker Solar Probe, whose flybys were used to lower its perihelion distance for solar observations, the ESA's BepiColombo, en route to Mercury, and the ESA's Solar Orbiter, which will make close observations of the polar regions of the Sun and is intended to answer questions about the Sun's heliosphere.

But suddenly, all this has changed; Venus is very much back in vogue, with circum-venusian space about to be invaded by a veritable international flotilla of spacecraft. After a long hiatus, Russia has resumed its historic interest in the planet. Its Venera-D mission has been in Phase A of development since the beginning of 2021. With a launch planned in 2029, this will be the first in a series of new Venera missions with an orbiter and a lander and possibly an aerial platform along with two long-lived surface stations provided by NASA which are designed to send data from the surface for at least two months.

Also, the Indian Space Research Organisation (ISRO), which had previously announced plans to launch a Venus orbiter in 2023 but because of work disruptions due to the global pandemic in 2020 and 2021 had to push back the schedule, is now expected to launch an orbiter called Shukraayan-1 (Venus Craft 1) in 2025. It is anticipated to carry two radars for surface and sub-surface surveys and a suite of atmospheric instruments.

The latest, and most ambitious, missions to Venus have been announced just as this book goes to press. In June 2021, after a hiatus of more than three decades since the launch of Magellan in

Artist's conception of DAVINCI probe descent stages.

DAVINCI probe Alpha.

Artist's concept of the Venus Emissivity, Radio Science, INSAR, Topography and Spectroscopy (VERITAS) spacecraft – a proposed mission for NASA's Discovery programme.

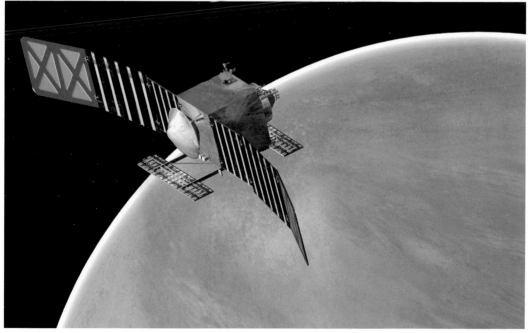

1989, NASA approved two Discovery class missions to Venus, which are tentatively scheduled for launch in 2028 or 2030. The first is DAVINCI (Deep Atmosphere Venus Investigation of Noble gases, Chemistry and Imaging), which will measure the composition of the atmosphere to understand how it formed and evolved, attempt to answer whether the planet ever had an ocean, and send back high-resolution images of the tesserae.[12] The second, VERITAS (Venus Emissivity, Radio Science, InSAR, Topography and Spectroscopy), will map the surface and produce 3D reconstructions of topography, in order to better understand Venus's geological history, the role of volatiles in crustal formation and whether active volcanism is still occurring.[13] Also, within a few days of NASA's announcement, the ESA approved EnVision, to be launched in 2031, which will carry two radars: an S-band synthetic aperture radar and another to probe the subsurface along with a near-infrared and UV spectral imaging spectrometer suite.[14]

Following the suggestion that cloud contrasts may be caused in part by the presence of micro-organisms, and reports in 2020 of traces of phosphine, interest in astrobiology investigations of Venus is emerging. A private concern, Rocket Lab, led by Peter Beck, as well as Breakthrough Initiatives, have announced probe missions to be launched in 2023 to sample clouds. This will be followed by a larger probe in 2024. The three radars on the ISRO Venus orbiter, VERITAS and EnVision can provide confirmation of active volcanism on Venus and insight into the presence of water in the past on Venus's surface, while DAVINCI will provide critical data on the altitude profiles of trace chemical species in the atmosphere below about 64 km along its descent to the surface in an attempt to better elucidate current habitability conditions. Even with all of these planned missions, the Venus era of exploration is just beginning. JAXA is planning to launch Akatsuki-2, a pair of two small orbiters, at the two co-linear Sun–Venus Lagrange points, L1 and L2, in the early 2030s.

Outstanding Questions

Venus, despite its superficial similarities to Earth, has obviously followed a very different evolutionary trajectory, and though important clues have been discovered, many mysteries still remain. Among the main questions are: did Venus have liquid water on the surface in its past? Did it always rotate slowly and backwards? If it had liquid water that survived for a billion years or so, did life originate? When did the surface warm up or was it always this hot? Did it ever have an magnetic field? Why does Venus have by far the most massive atmosphere of any terrestrial planet without a protective magnetic field?

Answering these evolutionary questions is challenging in the face of meagre data. For Earth and Mars, the geologic record offers some clues. For Venus, such clues have been hard to come by owing to the challenges of operating under the extreme conditions of heat and pressure and without any means of moving about the surface. At present concepts allowing greater mobility for exploring the surface and lower atmosphere are being developed, as well as high-temperature electronics, chemical sensors and mechanical drills.

In the early history of the solar system, after the Sun and planets formed from a protoplanetary disc some 4.6 billion years ago, the inner planets that emerged as smaller objects were gravitationally drawn together during a tempestuous and violent period of bombardment (called the Hadean eon in the geological history of Earth). Atmospheres on Venus, Earth and Mars would have formed by outgassing. Venus's, for reasons still not entirely clear, became extraordinarily massive over time (and unlike Mars's, did so even in the absence of a magnetic field). This massive atmosphere allowed only large impactors to reach the surface; the smaller ones disintegrated, producing massive amounts of dust. A study comparing what the rate of impacts would have been in the last 800 Myr if Venus had no atmosphere compared to in an atmosphere of current thickness shows a good fit with existing data.[15] This suggests that the atmosphere has probably remained about the same thickness over at least 800 Myr, but before this we have only rough estimates of rates of atmospheric escape into space and of outgassing from the surface.

Numerical modelling of Venus's climate suggests that if Venus was cool enough early on to have liquid water on its surface (which is by no means given), then it could have had shallow liquid water oceans for the first two or three Gyr, and perhaps retained them as recently as 800 Myr ago when cataclysmic volcanic resurfacing obliterated all earlier traces of venusian geological history.[16] For comparison, Mars, a more commonly considered astrobiology target than Venus, may have retained liquid oceans only until about 4.3 Gyr ago, as the rate at which its atmosphere was lost was very high.[17] Though by no means certain, it is at least possible that Venus had liquid water oceans longer than Mars did, and if so – seeing that life developed on Earth only several hundred Myr after the planet formed – the chance that life developed on Venus might be as great or even greater than that on Mars.

To stay with Mars for a moment, the evidence that supports the presence of liquid water on its surface early in its history includes degraded craters and residuals of flowing streams and lakes; while studies of atmospheric loss and geomorphology of Mars's surface is suggestive of ancient oceans, groundwater below the surface and even rainfall.[18] By contrast, we have no information about the early surface of Venus, as that has been obliterated globally by lava resurfacing. As we have said, we can't even say for sure if Venus ever had oceans, and so cannot know whether it ever developed a carbonate-silicate cycle such as that on Earth which locks carbon dioxide in calcium carbonate rocks and maintains the balance of carbon dioxide within a narrow range in the atmosphere. Clearly, there is nothing like that now; somehow along the way – we don't yet know how – Venus became unbalanced, with any carbon dioxide released into the atmosphere remaining there to form Venus's massive almost-entirely carbon-dioxide atmosphere.

With all its carbon dioxide remaining at large, the runaway greenhouse effect took off, and the planet's surface water (whatever remained) disappeared. At present, the surface is hot and dry as a bone, and drenched in a global cloud of sulphuric acid droplets. How long did it take Venus's oceans, assuming it once had them, to evaporate? Might there have been flowing rivers of water before the lava rivers emerged to run rampant over the surface? We know from past spacecraft results that deuterium is enriched over hydrogen in the Venus atmosphere at a level more than 150 times the ratio for ocean water on Earth. This may be evidence of past water on Venus. Additional indications come from the compositions of some minerals that form in the presence of water (felsic rocks). Improving our measures of the isotopic ratios of noble gases and surface rock composition (mineralogy) thus stand as key science goals of future missions.

The existence of Venus's massive atmosphere has other implications for the evolution of the planet. On Earth, the planet's primordial heat, produced by decay of radioactive elements, is largely dissipated through the processes of plate tectonics, through which heat escapes from the mantle by advection, mantle material is transported to the surface, then old crust is returned in subduction zones back to the mantle. On a planet without plate tectonics, the primoridal heat will remain trapped, presumably, until the mantle material becomes hot enough to push its way directly onto the surface. At that point the material will cool and the cycle will begin over again. In this scenario, Venus may have undergone planet-wide resurfacing events at regular intervals. The existence of Venus's thick greenhouse-gas-laden atmosphere will, of course, render the ability of Venus to radiate heat away into space far less efficient than in the case of Earth. Cooling will occur, but slowly, and episodes of volcanism during these resurfacing periods must remain long-lived. Indeed, it is doubtful whether the planet ever experiences periods without active volcanism. Venus, despite decades of spacecraft-based research, still offers more questions than answers.

NINE
LIFE ON VENUS

Of what nature are the inhabitants of Venus?
Do they resemble us in physical form? Are they
endowed with an intelligence analogous to ours?
Do they pass their life in pleasure, as Bernardin
de St. Pierre said, or, rather, are they so tormented
by the inclemency of their seasons that they have
no delicate perception, and are incapable of any
scientific or artistic attention? These are interesting
questions, to which we have no reply.

CAMILLE FLAMMARION (1842–1925),
POPULAR ASTRONOMY

Prior to the dawn of the spacecraft era, Venus was, like other
planets, a *tabula rasa*, on which humans could project their
dreams. The very clouds that made it radiant and alluring also
carefully guarded its secrets. Around the turn of the twentieth
century Mars, with its possible civilization of doomed canal-
builders, garnered the most attention as a potential abode of life.
But Venus and the Moon were far from being out of the running.
For instance, in 1912 the eminent French zoologist Edmond
Perrier, director of the Paris Musée de Jardins des Plantes, said in
a newspaper interview that ferns, insects and frogs as big as cows
were likely to exist on Venus. More definitively, the Nobel laureate
chemist Svante Arrhenius wrote in 1918:

The average temperature . . . is calculated to be about 47°C assuming the solar constant to be two calories per cubic centimetre. The humidity is probably about six times the average of that on the Earth, or three times that in the Congo . . . The atmosphere of Venus holds about as much water vapour 5 km above the surface as does the atmosphere of the Earth at the surface. We must conclude therefore that everything on Venus is dripping wet . . . A very great part of the surface of Venus is no doubt covered with swamps, corresponding to those on the Earth in which the coal deposits were formed, except that they are about 30°C warmer . . . The temperature . . . is not so high as to prevent a luxuriant vegetation. The constantly uniform climatic conditions which exist everywhere result in an entire absence of adaptation to changing exterior conditions. Only low forms of life are therefore represented, mostly no doubt belonging to the vegetable kingdom; and the organisms are nearly of the same kind all over the planet.

The venusian clouds that permanently hid the surface features of the planet also hid their own nature. At first, it was natural to assume that they were water vapour clouds, by analogy to those of Earth; and even as late as 1972, a survey of astronomers' opinions was quite evenly divided between those who believed the spectrum of Venus showed them to consist of water (24) and those who thought them to consist of something else (26).[1] The 'something

Giant frogs imagined by Edmond Perrier living on the surface of Venus (left) and a forest of ferns attracting birds, 1911.

else' included an impressive number of candidates, some quite exotic. Carbon polymers, silicates, even organic compounds such as polyoxymethylene (used on Earth as an engineering thermoplastic) all found adherents. In 1955 the cosmologist Fred Hoyle (of 'steady-state' fame) argued that the slow rotation, which he considered to be somewhat more than twenty days, could only be explained by friction of tides in some solution considerably more viscous than water. He suggested oil. In that case, he added, 'Venus is probably endowed beyond the dreams of the richest Texas oil-king.'[2] Knowing Fred Hoyle, his suggestion was probably tongue-in-cheek, but if true, it would imply the existence of widespread past or present biological life on Venus, since all oils found on Earth are of biological origin.

A Hellish Landscape

Mariner 2 established the extremely high temperature of the surface of the planet, and the Venera descent capsules measured conditions during their plunges through the atmosphere and finally from the surface itself. However, since all these spacecraft came down at low-latitude sites, it still seemed possible (if unlikely) that a more temperate regime might exist closer to the poles. Indeed, at one time it was even suspected that the bright cusp caps seen by visual observers might consist of snow or ice, like the polar caps present on Earth and Mars. Even as late as 1971, Willard F. Libby and Paul Corneil of UCLA were still arguing that polar seas might exist, that they were likely acidic and hence could not precipitate calcium carbonate, and that they might extend as far down as about 55° latitude.[3] While acknowledging the high temperatures found by spacecraft, they pointed out that at 100 atm the boiling point of water is 314°C (587 K), so that water might survive if the polar temperatures were significantly lower than those in the equatorial regions. (All the temperature measurements so far had been made in low-latitude regions.) Libby and another collaborator, Joseph Seckbach, went so far as to test these ideas in laboratory experiments in which they were able to show that, rather surprisingly, some algae, such as the red alga *Cyanidium caldarium*, thrive in a carbon dioxide atmosphere at 100 bar pressure and high temperatures.[4]

By 1974 James Hansen and J. W. Hovenier had achieved an important breakthrough in establishing that the venusian clouds did not consist of water vapour but of sulphuric acid droplets. By then, it had already been established that the surface temperature on Venus is hot, at least 460°C (860°F), though small variations occur due to topographic differences; the pressure changes by about 1 bar and the temperature by about 2°C (3.6°F) for every 160-metre difference in elevation near the surface. There is no significant difference between the poles and the equator, or between the night and day sides for that matter. So surface life was extremely unlikely. But there was another, more exotic possibility, suggested by Harold Morowitz and Carl Sagan, who in 1967 published a paper in the respected journal *Nature* with the provocative title, 'Life in the Clouds of Venus?'[5]

Life in the Clouds

In trying to determine which environments might be hospitable to life, historically we have always based our assumptions largely on our experience of life as we know it, that is, that on planet Earth. In the early telescopic era and even well into the twentieth century, the other planets, and even the Moon, were regarded as earthlike. More recently, what has been called the 'habitable zone' has been narrowed to planets like Earth at just the right distance from the Sun for temperatures to allow water to exist as a liquid on the surface. For this reason, since 1996 NASA has prioritized, at least for Mars missions, a philosophy of 'following the water', since the martian surface contains many features attesting to the existence of flowing water in the past. More recently, the broader theme of the search for life has been adopted as an element of the Planetary Science Decadal

Survey, and apart from Mars, other potential targets for astrobiology include Jupiter's moon Europa and Saturn's moon Enceladus. Venus, with its high surface temperatures and bone-dry conditions, has until recently been largely excluded from any discussion involving the presence of life, though as we shall see, this is now changing. Agreed, life as we know it certainly cannot exist on the surface at the present time. What about in the distant past?

Earth's biosphere remains the only place in the universe where we can study directly the emergence and survival of life. On Earth, the indispensable factors are the availability of surface and/or ground water over long geologic time periods. Therefore, in order to determine whether life could have emerged at some point in Venus's past, we need to determine the timeline and quantity of liquid water on early Venus, if it existed. As noted in Chapter Eight, the small number and relatively even distribution of impact craters superposed on the surface seen in Magellan images would appear to indicate an average surface age of only between 300 and 700 Myr old.[6] The first ~80–90 per cent of the geological record has been lost through resurfacing.[7] Thus any direct evidence of water on Venus's surface has long since been lost.

There is, however, evidence of a watery past in Venus's atmospheric composition, in spacecraft measurements with a mass spectrometer of the deuterium/hydrogen ratio found in Venus's atmosphere below the clouds. Deuterium is an isotope of hydrogen that contains a proton and a neutron, compared to ordinary hydrogen, which consists only of a proton.[8] Early on, the ratio of deuterium to hydrogen in the water on Earth and Venus must have been nearly the same. However, under the influence of solar ultraviolet radiation high in the atmosphere, water molecules dissociate into hydrogen and oxygen. Over time a significant amount of hydrogen and a lesser amount of deuterium (as it is twice as massive) escapes into space. According to measurements with the Pioneer Venus large probe mass spectrometer in 1978, the ratio of deuterium to hydrogen in Venus's atmosphere compared to Earth's is 150 – a result which can only be explained by Venus's having lost a massive amount of water. It has been generally assumed that it was liquid water, but a recent numerical simulation of early climate suggests it could have been steam with no possibility of condensation due to clouds keeping the planet very warm. More simulations are underway and may lead to contradictory results and hence future measurements are needed to settle this question. More recently, Venus Express measures have put the ratio above the clouds even higher, at 240 ± 25, possibly implying an even greater loss of water to space.[9] The escape of water is continuing apace, through the loss of hydrogen and oxygen ions, but a good estimate of the total loss per year is lacking.[10]

Modelling results imply that if Venus in fact once had an ocean of water, it would have taken at least 600 Myr to lose it through this photodissociative escape process.[11] It has also been argued that liquid water could have survived on Venus for 2 billion (Gyr) years or more, and that surface water and habitable conditions could have persisted, before simultaneous volcanic eruptions of large igneous provinces over the past few hundred million years led to the planet's current hellish state.[12] As noted earlier, on Mars by comparison liquid water oceans seem to have lasted less than 1 Gyr, as the rate at which the atmosphere was lost has been found to be quite high.[13] Since it appears that life on Earth arose after only several hundred Myr, the implications of this are that there is perhaps a greater likelihood of life having originated on Venus than on Mars.

Nevertheless, Mars remains the primary target of interest among astrobiologists, whereas Venus has ever since Mariner 2 been generally regarded as the epitome of inhospitability. In a recent astrobiology textbook, the past existence of liquid water on Venus is mentioned in only one chapter, and then only to be dismissed:

If Venus was habitable for about 2 Ga [billion years], and if life could have emerged in the planet, by comparison with the evolution of life on Earth, Venusian life forms could have reached quite a sophisticated level of evolution with the appearance of photosynthesis and even perhaps oxygenic photosynthesis. However, its closer proximity to the Sun resulted in what is termed a run-away greenhouse effect . . . that resulted in Venus becoming inhabitable to such an extent that temperatures > 350 °C at the surface, an atmosphere rich in sulphuric acid, together with volcanic resurfacing have most likely erased any trace of potential past life.[14]

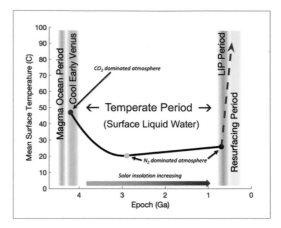

Evolution of water on Venus. Data points represent mean surface temperatures at 1 bar of atmospheric pressure. The red dashed arrow represents the transition to a runaway greenhouse atmosphere.

The possibility that life – perhaps in the form of tiny nanoscale bacteria thriving on micro quantities of nutrients, adapted to extremes of atmospheric pressure, temperature, acidity and radiation – could have migrated to a more hospitable cloud zone above the surface was not considered, and was presumably regarded as outlandish.

Apart from the preliminary suggestion of Morowitz and Sagan that the clouds of Venus might harbour life, the literature on the subject has been rather sparse, as well as speculative and devoid of any substantive arguments, until recently. However, several early papers invoked (very tentative) microbial explanations for the ultraviolet absorption in the clouds. In 1975 Bruce Hapke and Robert Nelson at the University of Pittsburgh, after showing that the decrease of reflectivity in the near UV could be explained if the clouds contained particles of elemental sulphur in addition to sulphuric acid, concluded with a tantalizing comment:

Finally, we cannot resist pointing out that many examples
of anaerobic, terrestrial organisms are known in which
the reduction or oxidation of various forms of sulphur are
important sources of energy in their metabolisms . . . If these
microorganisms do not find conditions in the clouds totally
inimical, their effect on the energy balance of the planet may not
be negligible.[15]

Two years later, Mikio Shimizu of the University of Tokyo argued
that the UV absorbers were likely carbon suboxide polymers and
fine graphite grains formed in Venus's atmosphere by irradiation
of carbon monoxide by solar ultraviolet radiation, and added that
these might have been the 'precursors of the living system on the
primitive Earth', so that, 'symbolically speaking, we may be seeing
our "ancestors" every time we are looking at the UV photo (bright
and dark stripes) of the Venus clouds'.[16]

Another provocative idea was offered by the early pioneer of
UV Venus imaging Charles Boyer, who after retiring moved back
to Toulouse and continued to collaborate with Henri Camichel
and other Pic du Midi astronomers. In 1986, in the French popular
astronomy journal L'Astronomie, he imaginatively envisioned a vast
sheet of photosynthesizing organisms, behaving like the algal
blooms in our oceans, growing in size until the available nutrients
or water are depleted, then dying out, all over a matter of a few days.
He thought that this would explain the observed variations in the
zonal flow indicated by the measured cloud motions. He wrote,

This layer floats at an altitude of 60 km. One can validly
assume that this life is sustained by reactions analogous to
photosynthesis, and to absorption by some kind of chlorophyll.
The part of the sheet in the sun restores its reserves while the
part in the dark exhausts them. Since the latter lasts about
48 hours at the cloud tops, when they emerge from the dark,

the microorganisms of the sheet are ravenous. Some three or four hours before reaching the sunlight, they anticipate what lies ahead by a rudimentary nervous system, and begin their rush toward the beneficent sun. As soon as they arrive, this acceleration ceases. Now a deceleration begins.[17]

Boyer may be forgiven a bit of poetic licence, for there is no need to invoke a nervous system to explain this behaviour. It would be expected to occur as a natural response to a solar thermal tide in the atmosphere.

Fanciful, perhaps, though the idea was later taken up by the American planetary scientist David Grinspoon, who also speculated that the 'unknown ultraviolet absorber is a photosynthetic pigment'.[18] Like Boyer, he did not suggest any specific terrestrial analogues to the hypothesized photosynthesizing organisms. Recently, Boyer's idea has received support – photosynthesis is indeed possible in the clouds, according to Rakesh Mogul and colleagues, not just on day side, but even on the night side due to the emission from the lower atmosphere and surface.

The evolutionary history of life on Venus may, at least in the early stages, have resembled that on Earth. What do we know about the origin of life on Earth? It is possible that the panspermia idea is true – that life exists throughout the universe and is seeded from world to world by space dust, meteoroids, comets and the like. In that case life would presumably have been seeded on both Earth and Venus. On the other hand, if life did have a special origin on Earth, then it seems that Charles Darwin may have had the right general idea when he speculated a century and a half ago that life might have begun in a 'warm little pond with all sorts of ammonia and phosphoric salts, – light, heat, electricity present'.[19] Darwin believed that life began near the surface. However, the discovery in 1979 of hydrothermal vents ('white smokers' and 'black smokers') at mid-ocean ridges, where early life would have been protected by the deep ocean from hazards such as intense meteoritic surface bombardment during the Hadean eon (4.6–4.0 Gyr ago), led many researchers to believe that deep-ocean locations probably harboured Earth's earliest life forms.[20] This idea still has supporters, though in recent years the pendulum has begun to swing back again. The problem with hydrothermal vents is that, though essential to life, water is also highly destructive to the core components of cells such as DNA and proteins, and their formation would have been more likely in land environments that were alternately wet and dry, that is, looking more like Darwin's 'warm little pond' than deep-ocean hydrothermal vents.[21]

Be this as it may, hot springs and hydrothermal vents are still very interesting places, and provide habitats to many exotic organisms, including 'extremophiles' able to survive and even thrive under conditions of extremely high temperature, pressure and acidity. The most extreme extremophiles are archaea – the oldest of any still-extant lineage of organisms on Earth – which include sulphur-eating bacteria that use inorganic sulphur compounds as electron acceptors instead of oxygen and methanogens which produce methane as a metabolic by-product.[22] Some survive in temperatures above 75°, 90° and even 100°C, the temperature of boiling water. Among these, the archaea *Pyrolobus fumarii*, extracted from the wall of a black smoker in the Atlantic in 1997, can grow at an astounding temperature of 113°C, probably close to the limit of what is possible, since at still higher temperatures almost all biomolecules will fall to pieces within seconds. Microorganisms are known that can grow in the deep ocean at pressures between 300 and 700 bar and temperatures between 2 and 4°C, while others (such as the archaea *Sulfolobus*) are both acidophilic and thermophilic and thrive in hot springs and solfatara fields, where they happily convert hydrogen sulphide to sulphuric acid. Dirk Schulze-Makuch and colleagues proposed acid-resistant bacteria as possible habitants in the clouds and argued for a balloon mission.

The following are extremophile categories relevant to Venus:

Acidophiles – organisms immune to acidic environments with pH levels of 3 and below.
Barophiles – organisms surviving in high- and low-pressure environments.
Psychrophiles – organisms capable of living in very cold conditions between −20°C and +10°C.
Radiophiles – organisms that thrive in conditions with high levels of UV and ionizing radiation.
Thermophiles – organisms that survive at high temperature, including *hyperthermophiles*, which can withstand temperatures above 75°C.
Xerophiles – organisms that live in extremely dry conditions.

In addition, some microorganisms are able to survive exposure to solar wind and cosmic rays, as the Long Duration Exposure Facility experiments showed, making it nearly impossible to render a spacecraft 'clean' with currently used cleaning agents (indeed, some species actually thrive on the cleaning agents). Some of these may be able to survive for at least limited periods in the upper cloud layer or even in the mesosphere of Venus, where radiation exposure is so deadly.

However life gets started, we know that once it does life proves remarkably persistent and able to adapt to an incredible diversity of conditions. This is certainly so on Earth, and presumably it would have been so on Venus. New research indicates that oceans and habitable conditions could have persisted on Venus's surface for several Gyr. However, eventually, probably owing to simultaneous volcanic eruptions associated with cataclysmic resurfacing events, Venus lost its oceans and evolved into its current stage of possessing a massive nearly-all-carbon-dioxide atmosphere and a runaway greenhouse effect. Conditions on the surface of Venus now could hardly be more inhospitable to life as we know it or can even imagine it. The only way out, if there was one, would have been upwards, with microorganisms hitching rides on surface winds and carried into the sulphuric acid clouds by topography-induced gravity waves such as those recently detected by the Akatsuki orbiter. On Earth viable bacteria have been detected as high as the sulphuric acid aerosol layer (or Junge layer) in the stratosphere, where the chemistry is comparable to that of Venus's atmosphere.[23] However, viable microbes recovered from stratospheric air samples are very sparse and largely limited to metabolically inactive forms that are able to survive extended desiccation and UV exposure. Cell densities in the stratosphere can reach some 10^5 cells/m^3, compared to some 10^{11} cells/m^3 in the cloud-forming regions of the lower troposphere. Interestingly, these tropospheric values are similar to maximum cell density estimates for Venus's lower clouds of 10^8–10^{10} cells/m^3 as calculated by co-author Limaye.[24]

Compared to Earth, the conditions in Venus's clouds present many extremes of desiccation, acidity and quite possibly low availability of nutrients. As inferred from particle sizes and densities, the cloud-based microorganisms hypothesized by Limaye and colleagues would be recycled between lower and upper extremes of the cloud layer through the merging and division of cloud droplets, which theoretically would retain access to water and nutrients through bulk mixing. However, the acidity within the aerosols on Venus is likely much greater than the reported pH range of 1.5 to 0.5 estimated for the clouds between 48 and 65 km.[25] On the other hand, protective membranes such as those found in spores may allow them to survive in high acidity. Also, spores may act as active cloud condensation nuclei. Sara Seager of the Massachusetts Institute of Technology (MIT) and her colleagues have proposed a vertical life cycle for microorganisms across the cloud layers wherein the microorganisms fall out from the lower cloud as they grow larger, then form spores below the cloud layer.[26]

A great deal still needs to be worked out, but at least we can begin to visualize what the 'inhabitants' of Venus might look like, if there are any. They would look nothing like the multicellular creatures imagined by Edmond Perrier. Instead, they would probably resemble sulphur-reducing archaea or bacteria (no specific terrestrial species is implied; Venus's microbial life would have followed its own evolutionary pathways). The clouds on Venus lie between about 47 km (~110°C, 2 bar) and ~70 km (−45°C, 40 mb). At these altitudes, the amount of water vapour is very low (5 ppm at the top to ~50 ppm at the bottom based on a few measurements or inferences of remote data and could be variable), but a haze of smaller particles extends high into the atmosphere to about 90 km. More water in liquid form is in the droplets of sulphuric acid, which make up the clouds. The clouds are layered, and contain particles, assumed to be spherical in shape, ranging in size from about 400 to 2,500 nm, as well as larger particles, perhaps as large as 20,000 nm in size, which may not be spherical. According to current models, microorganisms would concentrate near the bottom of the cloud layer, at about 50 km, where the amount of water in sulphuric acid droplets is relatively high. Acidity in the droplets could be ameliorated by the presence of some impurities in the form of salts, as has been suggested from observations of the glory feature seen in the optical images or polarization data and surmised from chemical models and mass spectrometer data. Local circulation might carry them to higher altitudes near the cloud tops at 70 km, where they would succumb to UV desiccation and also ionizing radiation perhaps decaying photochemically, dropping back down through the clouds, and surviving by forming spores below the cloud layer. The point is that the microorganisms in their various forms could live out their entire life cycles along a vertical dimension, without ever getting near the inhospitable surface at all.

Might such microorganisms be the still-mysterious UV absorbers? The measurements of sunlight at increasing depth from

Composite (left to right) of Akatsuki images of Venus taken through 283 nm, 365 nm and 900 nm filters.

Venera and Pioneer probes showed that much of the absorption of sunlight takes place in the clouds, with only a small fraction straggling through to reach the surface. At noon, only about 20 per cent of the sunlight reaches the surface of Venus, so despite the fact that Venus is closer to the Sun, the intensity of sunlight at the surface is comparable to that on a particularly gloomy, cloudy day on Earth. The Sun itself is not distinguishable as a bright spot. Also, as is known from the Venera lander images, the sky is orange. About half of the absorption by the clouds takes place in wavelengths shorter than 400 nm, corresponding to the spectral region in which the blue–violet contrast features are faintly visible to some visual observers, and strongly registered in UV photographs.[27] Studies of the motions of these contrast features have revealed the nature of

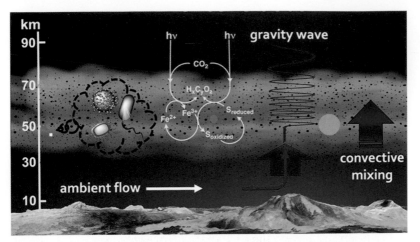

km
90
70
50
30
10

hν   hν   gravity wave

CO_2

$H_2C_nO_x$

Fe^{2+} Fe^{3+}

$S_{reduced}$

$S_{oxidized}$

convective
mixing

ambient flow →

Schematic view of possible
microorganisms in Venus's
clouds.

Venus's atmospheric winds. The large probe on the Pioneer Venus
Multiprobe mission in 1978 showed that much of the ultraviolet
light is absorbed completely by the time it reaches 57 km. VeGa 1
and VeGa 2 lander measurements added further data, using a xenon
light (the probes descended at midnight) to show that the ultraviolet
absorbers remain present all the way to the bottom of the clouds
at 47 km. The cloud layers where the UV absorption takes place
correspond to the presumed habit of venusian microorganisms, if
they exist.

Whatever they are, the ultraviolet absorbers are peculiar, in
that it is hard to understand why any contrasts should exist in the
venusian cloud cover at all. Except for their absorption properties,
the dark and bright areas in the clouds show no discernible
differences in physical properties. Why, then, aren't the absorbers
uniformly mixed? (To recall a similar case, the dark features seen on
Jupiter after the impact of the Shoemaker-Levy 9 fragments in July
1994 disappeared within about a year.) It follows, with fast winds
and propagating gravity waves and other waves of different scales
mixing momentum and heat throughout the venusian atmosphere,
that the absorbers ought to be be mixed globally, unless there are
localized sources and sinks.

But microorganisms fit these features of the UV absorbers as well. Indeed, it has been shown that terrestrial bacteria and proteins absorb in the same ultraviolet wavelength region as the UV absorbers on Venus. In particular, the absorption spectrum of *Thiobacillus ferrooxidans*, a gram-negative highly acidophilic (pH 1.5 to 2.0) autotrophic bacterium that obtains its energy through the oxidation of ferrous iron or reduced inorganic sulphur or sulphur compounds, is very similar to that exhibited by Venus. Indeed, an iron–sulphur-based metabolism has been proposed for microorganisms concentrated in the lower cloud layer that could absorb short wavelengths.[28] Perhaps Boyer's basic idea will turn out to be right after all: the observed contrasts can be explained as microorganism colonies which grow in size until the available nutrients or water are depleted and decay over time, analogous to algal blooms in oceans. Colonies may then appear elsewhere as dead organisms are scavenged.

If water had indeed been able to exist on Venus's surface for a long time, as current models of the planet's evolution suggest, then diverse microorganisms could have evolved and adapted to increasingly extreme conditions and survived for the last few Myr in the clouds, recycling between forming spores through desiccation and coming alive as soon as liquid water in the cloud droplets could be harvested. This may sound far-fetched, but the only way to find out if this is happening will be by sampling the entire cloud layer at different locations and at different times of day.

Looking for Biosignatures on Venus

In order to demonstrate the existence of life in Venus's clouds, we need a biosignature of some sort, something – a chemical compound, isotope or cellular component – that furnishes definitive evidence of the presence of biological processes. Among these biosignatures are gases such as phosphine, methane and ammonia.[29] Phosphine (PH_3), whose production requires a reducing atmosphere, has been of greatest interest recently following a September 2020 report of detection of phosphine by a team led by Jane Greaves of Cardiff University, using the Atacama Large Millimeter/submillimeter Array (ALMA) in Chile.[30] They found the average phosphine levels across the planet to be about 7 parts per billion. The result was a surprise, since phosphine wouldn't be expected to survive very long in Venus's atmosphere; its existence implied that it was somehow being replenished – perhaps by microorganisms. The Greaves result was almost immediately challenged, and on going back over their data, Greaves and her colleagues found that their original data set contained a spurious signal that could have affected the result. On redoing part of their work using both ALMA and the 15-metre James Clerk Maxwell telescope in Hawaii, however, they again found the phosphine signal, but at a reduced average level of only 1 part per billion, though with occasional peaks of as much as 5 ppb, suggesting the abundance might vary over different parts of the planet. Both the original paper and the partial retraction generated a great deal of interest, and a number of other scientific papers on this topic followed. Some researchers claimed to have found no evidence of phosphine at all, but others reaffirmed its presence when they reanalysed old Pioneer Venus observations. Subsequently it was proposed that rather than being produced by microorganisms, the phosphine could form when phosphorus compounds from deep within the venusian mantle are released in Krakatoa-like eruptions high into the clouds of Venus, and combine with sulphuric acid to form phosphine. Or, the phosphides could be brought in by impactors. At the moment, all of these explanations remain speculative, and we don't really know for sure whether Venus has volcanism at the present time, or the rate of volcanic outgassing or the likely composition of volcanic gases. Nor is it clear, even if the phosphorous compounds did reach Venus's clouds, whether their interactions with sulphuric acid would produce phosphine. Clearly, additional spacecraft observations are needed.

Whatever the UV absorbers are, they are able to alter the albedo of the clouds enough to change the planet's weather over time. In contrast to Earth, where most of the energy from the Sun is absorbed at ground level, on Venus most of the heat is deposited in the clouds. Observations documenting more than a decade of UV observations of the planet from instruments aboard the planetary probes Venus Express, Akatsuki and MESSENGER, as well as the Hubble Space Telescope, have shown that Venus's albedo in the UV (365 nm) decreased on a global scale by about half between 2006 and 2017, before beginning to rebound. These changes produced large variations in the amount of solar energy absorbed by the clouds and even affected the zonal winds at the cloud-top level. The variations might involve sulphur dioxide gas abundance above the clouds, but also, since the lowest albedo recorded corresponded to the sunspot maximum, there appear to be links with the solar cycle and consequent galactic cosmic-ray variations.[31] It is all rather mysterious, but clearly something very interesting is going on. The Venus clouds remain the most puzzling aspect of Venus given the little data that we have.

After long being written off as a subject of investigation, Venus has been rediscovered! The bevy of spacecraft missions recently approved will hopefully be able to shed light on this and other questions about the planet (as well as, no doubt, discovering many more). A special Venus collection has been published by Astrobiology journal in the October 2021 issue. It deals with the habitability of the Venus cloud layer following the first ever workshop hosted by the Space Research Institute in Moscow, and organized by the Russia-U.S. Joint Science Definition Team for Venera-D mission. A second workshop is planned for November–December 2021 and more research results are anticipated. The main questions for which we would like to find answers are:

1. Did Venus always spin backwards?
2. Are there active volcanoes today?
3. How long has the current cloud cover existed?
4. What are the absorbers of sunlight in the Venus atmosphere and clouds?
5. How has the climate evolved and when did Venus become so hot?
6. How much liquid water was present in the past on Venus?
7. Where did the water come from?
8. Did life form on Venus if liquid water was present for most of its history as modelling suggests, and if so, did it evolve and survive?
9. How did the venusian atmospheric circulation become super rotating?
10. Did Venus ever have a magnetic field, and if so, how was it lost?
11. How much atmosphere is being lost due to solar wind?

The lovely Evening or Morning Star, known from times immemorial as without comparison the most magnificent star of our sky – the Babylonian 'Queen of Heaven', which cast like a shadow the circle of its enchantment on lovers and poets and children going home to their mothers – remains, even in the spacecraft era, one of the most intriguing of all the worlds in the solar system. Much as we have learned about it, there is still much that we do not yet understand. Meanwhile, as the late Richard Baum, one of the leading amateur students of the planet, used to like to say, 'The mystery endures.'

Viva la mystery!

TEN

OBSERVING VENUS

B ecause of its brilliance, Venus is an irresistible object for the
telescopic observer. A good pair of binoculars will readily
reveal the crescent phase, though they need to be stably mounted
for the purpose. The sight of the planet's phase can still astonish
with its visual proof of the Copernican system, and is hard to
forget. Thus, the great American astronomer Edward Emerson
Barnard recalled that his youthful sighting of the planet's phase
'made a more profound and pleasing impression' than did his later
discovery of the fifth satellite of Jupiter, to which his colleague
S. W. Burnham would add, 'Such excitement comes but once in a
lifetime, although the enthusiasm and interest in the subject may
never be abated.'[1]

Apart from the phase, however, the usual appearance can only
be described as disappointing. Not for nothing has Venus earned its
reputation for being an unrewarding object for the visual telescopic
observer. Observing against a dark sky when Venus is glaring like
a flare is clearly pointless, but early twilight sometimes gives good
views. Most systematic observers prefer daylight conditions, with
Venus higher in the sky and its overpowering brilliance mitigated.
When Venus can be seen with the naked eye, it is possible to point
the telescope to it directly; when it is close to the Sun, setting circles
or a mount with GoTo capability are needed to safely point it to the
right position. (Please take care when sweeping near the Sun.)

Though illusory markings, including the hub-and-spoke system seen by Lowell and a few others, are common, especially in small telescopes used at low powers, with practice and determination observers may detect something of the actual markings in the cloud deck. The French astronomer Audouin Dollfus suggested that the illusory markings tend to disappear and the real markings to appear with higher magnifications of about 400 ×; he personally usually used a magnification of 900 × on Venus. They are most pronounced along the terminator and recall those recorded by Bianchini almost three centuries ago. Their visibility depends on the seeing conditions and, no doubt, on the observer's sensitivity to the blue and violet part of the spectrum. Colour filters are useful. Red (Wratten 23A and 25) and orange (Wratten 21) enhance the sharpness of the limb and terminator, but show little detail on the disc. Low-contrast markings sometimes can be made out in yellow light (Wratten 8, 12 and 15).

Venus in the Ultraviolet

We now know that the most prominent markings on Venus are the dark patches indicating the presence of the still-mysterious UV absorbers first clearly revealed in Frank Ross's photographs from the 1920s. The question thus arises to what extent the markings seen by visual observers, which proved so unhelpful when it came to working out the rotation period of the solid planet, correspond with these – or indeed have any reality at all.

In the third edition of *The Planet Venus*, published in 1961 and the last word on the subject in the pre-spacecraft era, the then director of the Mercury and Venus Section of the British Astronomical Association, Patrick Moore, summed up three centuries of visual observers' effort to determine the planet's rotation period:

In the main, we have to admit defeat – for the moment. The visual observer has fought a losing battle; certainly my own hundreds of drawings, made between 1934 and the present time, shed no light on the subject, and this has also been the frustrating experience of many far better observers. Even those who have tackled the problem by means of the photographic plate, the spectroscope, and the radio telescope cannot yet say how long Venus takes to spin once upon its axis. Whether we will find out in the near future remains to be seen.[2]

Those were the best of times and the worst of times for Venus studies. They were the best of times in that amateur work was still of value; they were the worst of times since so little was still definitely known. Moore, with his tremendous energy and enthusiasm, did a great deal to revive amateur interest in Venus at the time. Another leading BAA observer was Richard Baum, of Chester, England, who had begun to make observations of Venus with a borrowed 8-centimetre refractor as far back as 1950, when he was only twenty and just home from service in the Royal Air Force. At first he had the usual experience of making out nothing at all. However, on 19 February 1951, he began to suspect a dusky sub-solar spot, and throughout the apparition, especially in March and April, had a persistent impression not only of the dusky spot but of 'an extremely curious system of dark longitudinal streaks', which he identified with Percival Lowell's infamous hub-and-spoke system.

Moore, just coming into prominence at the time, gave a talk on Venus at the Grosvenor Museum in Chester in August 1953, at which Baum was in the audience. He equated Baum's drawings with Lowell's; someone suggested that there was a greater resemblance to the markings in Ross's UV photographs, at which point C. D. Reid, an atomic energy industry employee as well as an amateur astronomer, wondered whether different observers were likely to have varying sensitivities to the ultraviolet. As Reid

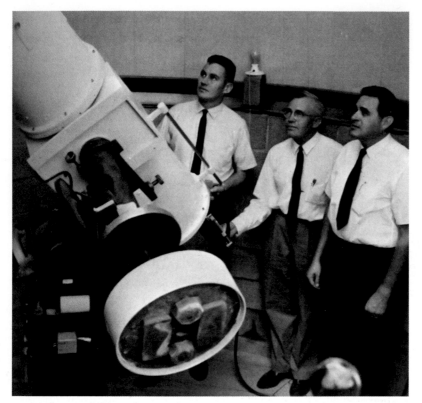

Bradford A. Smith, Clyde Tombaugh and Jim Robinson posing in Smith's back garden in 1964 with the 30.5-cm reflector then being used to photograph the planets at New Mexico State University. The group, under Tombaugh's supervision, later moved to Tortugas Mountain (known locally as 'A' Mountain), using a 61-cm reflector for its planetary patrol programme from 1969 until the observatory was closed in 1993.

himself had access to a very good spectrometer, he suggested that the experiment be tried. The following morning in a blacked-out room in Reid's home, the UV sensitivity of Reid, Baum, Moore and J. B. Hutchings was tested, but the results were not announced until April 1954, when the men were again brought together for an astronomical meeting in Chester.

On the evening of 21 April, they observed Venus through Reid's 23-centimetre reflector, and each made a sketch of the planet. Baum's showed a Y-shaped marking, Moore's only vague and fuzzy patches. After the observation, Reid summed up the results of the experiment performed in the previous year. Baum had demonstrated a strong UV facility, down to about 320 nm; Reid and Hutchings were intermediate at about 370 nm; Moore performed worst, at about 400 nm. Thus it seemed that differences in UV sensitivity might explain the dissimilarity in observers' views of Venus.

Baum was encouraged, and continued to hold forth for the reality of the markings he saw on the planet. In 1960 he submitted some of his drawings to Bradford A. Smith, who was then imaging Venus in the UV at New Mexico State University. Though Smith found that detail in Baum's (and one other observer's) drawings resembled that recorded in his UV photographs, these were the exceptions, and in general he found:

> Our photographic studies have shown that the contrast of this band structure is highly dependent upon wavelength, and that readily apparent detail in the ultraviolet may be traced, with difficulty, into the blue–violet and blue regions [of the visible spectrum], disappearing completely in the green and beyond. As a result of these studies, I find it most remarkable that you are able to see this structure at all in the visual regions. Parenthetically I might remark that the drawings sent to me by other visual observers have borne little or no resemblance to the ultraviolet photographs, and I am forced to conclude that there are but few reliable observers of the planet Venus.[3]

In light of the possibility (admittedly rather remote) that these markings might consist of wind-hurtled colonies of venusian microbes, they have certainly become more intriguing. But can a visual observer even hope to make them out?

From Akatsuki and MESSENGER imaging, the markings appear most prominently at UV wavelengths of 283 and 365 nm.[4] They tend to fade at wavelengths above about 370 nm, though as Ross showed in his coloured-filter images in the 1920s, they can still be weakly detected in the violet and blue between 400 and 450 nm. Faint details can also be seen at longer wavelengths, as known from Mariner 10 orange-filter images and MESSENGER images.

The would-be observer of the UV features should search out Venus high in a daylight sky, and use a neutral-density filter. Reasonably high magnifications are needed. Observers at higher elevations are also favoured, as UV radiation transmitted by the atmosphere increases by 10 to 12 per cent per 1,000 m. Those with reflectors have an advantage over those with refractors, unless the refractors have lens elements made from fused quartz or quartz and fluorite. This is because, while aluminium mirror coatings are highly reflective of the UV, the flint glass in ordinary achromatic lenses absorbs wavelengths shorter than 350 nm.

The most important determinant of ability to visualize the UV markings is the crystalline lens of the human eye. The crystalline lens of the adult eye absorbs almost all incident energy below a wavelength of 400 nm, but in newborns and children, a small transmission window occurs at around 320 nm, remaining until age ten or so. At age eight, peak absorption is 365 nm, but falls off to 450 nm by age 65; indeed, for older individuals, the crystalline lens becomes so yellowish as to act like a yellow (or even an orange) filter.[5] As with any physiological parameter, individual differences are significant. There are significant variations around the mean,

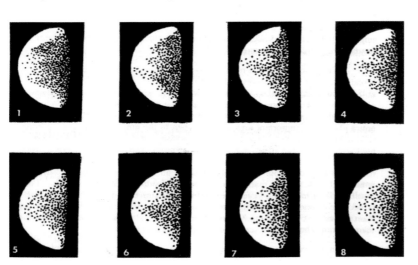

Drawings of Venus at eastern elongation, March 1988, using a 15-cm reflector, ×100, by Ewen A. Whitaker (Tucson, Arizona). Following cataract removal, Whitaker found he could detect UV light visually down to 317 nm, and readily observed the UV markings on the planet using a filter transmitting wavelengths shorter than 410 nm.

Bland, lovely Venus. This image was taken by William Leatherbarrow using a 30.5-cm Maksutov-Cassegrain and a 742 nm pass filter at Sheffield, England, on 7 May 2020, and corresponds to the usual view in visible and near-infrared wavelengths.

and in post-mortem studies, one 24-year-old had a pronounced window at 330 nm. It follows that in general, young people are likely to be the best observers of the UV markings.

An exception is older individuals who have had cataract surgery, and here an experiment by the well-known lunar scientist Ewen A. Whitaker is worth mentioning. At age 66, Whitaker had his cataracts removed, but in his right eye had a plastic lens implanted without the usual 'minus UV coating'. This allowed him to see into the UV to a wavelength of about 317 nm. He made a number of observations of Venus with a 15-centimetre reflector, using a neutral-tinted filter, and found the UV features strikingly evident. Indeed, he wrote that they appeared 'very similar to the naked-eye contrast of the lunar maria'.[6]

Imaging Venus

The possible detection of the UV features by visual observers apart, the greatest scope, of course, lies open to imagers. Charge-coupled-device (CCD) imaging with blue or violet filters (Wratten 38A with telescopes of less than 15 cm, and Wratten 47 in larger apertures) will show the UV features readily, of which the most persistent is the planetary-scale dark Y-feature, which rotates round the planet with a period of four to five days. (However, it often changes markedly from one apparition to the next; this is something worthy of systematic investigation.) As noted above, reflecting telescopes are best, since aluminium coatings are highly reflective in short wavelengths. Otherwise, refractors are best avoided except those with specially developed lenses with elements of fused quartz or

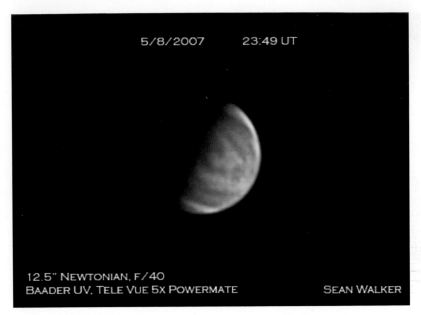

5/8/2007   23:49 UT

12.5" NEWTONIAN, F/40
BAADER UV, TELE VUE 5X POWERMATE   SEAN WALKER

An amateur image, taken with a Baader UV filter and a 32.75-cm Newtonian reflector, showing complex details in the clouds.

quartz and fluorite. In ordinary achromats, the flint glass element strongly absorbs UV light and is virtually opaque to wavelengths shorter than 350 nm.[7]

Both visual observers and imagers should be on the lookout for occasional bluntings and extensions of the cusps (the latter, of course, are always seen when Venus is within about 2° of the Sun near inferior conjunction). Occasional defects or bumps in the terminator appear; an indentation on the terminator just below the southern cusp, first seen by Giovanni Schiaparelli in 1877, is conspicuous at times, leading to the impression that the cusp cap is divided into two spots. Such features are thought to be due to particularly elevated or depressed cloud formations. Also, the magnitude of the 'Schroeter effect', discovered by Johann Hieronymus Schroeter in 1793, in which Venus appears exactly half phase several days from the theoretical time of dichotomy due to scattering of sunlight in the atmosphere of Venus above the cloud deck, is worth noting.

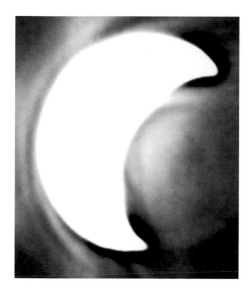

For this image taken in May 2004, Christophe Pellier of Bruz, France, combined images taken with a 36-cm Schmidt–Cassegrain telescope, a webcam, and a 1,000 nm infrared filter to successfully register the thermally glowing night side. Some cooler highland regions appear faintly as dark spots in the dusky area to the right of the drastically overexposed illuminated portion.

The bright polar caps or cusp caps, often surrounded by dusky collars, were first reported in 1813 by Gruithuisen with only a 60-millimetre refractor. Long believed to be illusory, they have been identified with the bright polar cloud swirls imaged by Mariner 10, Pioneer Venus and other spacecraft. The dusky collars correspond to a band of cold clouds that surround the poles of Venus at latitudes of about 70°. The polar caps are usually seen as bright amorphous features, although at times they may appear to break up into two or more discrete bright spots, and they sometimes undergo rapid changes in size and brightness. Also, the caps are usually asymmetric, with the southern cap typically being recorded as larger and brighter than the northern one.

An Astronomical 'Loch Ness': The Ashen Light

As we now know, the surface of Venus is intensely hot, and radiates in the infrared at wavelengths of 1,000–2,000 nm. Since the atmosphere of Venus is partially transparent in these wavelengths, in principle the planet's surface topography can be imaged through the clouds, cooler regions appearing darker than their surroundings.[8] CCD imagery even by amateurs now routinely detects features on the surface that are invisible to the human eye. (It turns out that in his pioneering work with colour filters, Ross came close to revealing these features; his IR filter only admitted wavelengths longer than 900 nm, so he just missed the cut-off.) It is important to emphasize that the thermal emission of the night side is well beyond detection by ordinary means, and has nothing to do with the so-called 'Ashen Light', a pale illumination of the night side which has been reported from time to time by visual observers going back to Giovanni Riccioli in 1643. The latter seems to be an illusion, though it may be that the last word has not yet been said.[9]

Though nowadays, with spacecraft circling the planet, the field for scientifically useful amateur work is admittedly rather limited, there can be no doubt that Venus remains one of the most intriguing of all worlds. It will always be well worth visiting and pondering as a tourist. As Richard Baum once wrote, 'Venus usually gets a bad press in the descriptive literature. Regular devotees know it is otherwise. It is a beautiful object that fascinates as it frustrates.'[10]

It is well worth another look.

Appendix 1:

Venus Data

(Source: Planetary Fact Sheet, National Space Science Data Archive, https://nssdc.gsfc.nasa.gov/planetary/factsheet/venusfact.html)

Quantity	Venus	Earth	Ratio (Venus–Earth)
Mass (10^{24} kg)	4.8675	5.9724	0.815
Volume (10^{10} km³)	92.843	108.321	0.857
Equatorial radius (km)	6051.8	6378.1	0.949
Polar radius (km)	6051.8	6356.8	0.952
Volumetric mean radius (km)	6051.8	6371	0.95
Ellipticity (flattening)	0	0.00335	0
Mean density (kg/m³)	5243	5514	0.951
Surface gravity (eq.) (m/s²)	8.87	9.8	0.905
Surface acceleration (eq.) (m/s²)	8.87	9.78	0.907
Escape velocity (km/s)	10.36	11.19	0.926
GM (x 10^6 km³/s²)	0.32486	0.3986	0.815
Bond albedo	0.77	0.306	2.52
Geometric albedo	0.689	0.434	1.59
v-band magnitude v(1,0)	−4.38	−3.99	–
Solar irradiance (W/m²)	2601.3	1361	1.911
Black-body temperature (K)	226.6	254	0.892
Topographic range (km)	13	20	0.65
Moment of inertia (I/MR^2)	0.33	0.3308	0.998
J_2 (x 10^{-6})	4.458	1082.63	0.004

Orbital Data

Semimajor axis (AU)	0.72333199
Orbital eccentricity	0.00677323
Orbital inclination (deg)	3.39471
Longitude of ascending node (deg)	76.68069
Longitude of perihelion (deg)	131.53298
Mean longitude (deg)	181.97973

AU = 149

North Pole of Rotation

Right ascension: 272.76°

Declination: 67.16 deg

Reference date: 12:00 UT 1 Jan 2000 (JD 2451545.0)

Venus Observational Parameters

Distance from Earth	
Minimum (10^6 km)	38.2
Maximum (10^6 km)	261
Apparent diameter from Earth	
Maximum (seconds of arc)	66
Minimum (seconds of arc)	9.7
Maximum visual magnitude	−4.8
Mean values at inferior conjunction with Earth	
Distance from Earth (10^6 km)	41.44
Apparent diameter (seconds of arc)	60.2

Venus atmosphere	
Surface pressure	92 bar
Surface density	~65. kg/m^3
Scale height	15.9 km
Total mass of atmosphere	~4.8 × 10^{20} kg
Average temperature	737 K (464°C)
Diurnal temperature range at surface	~0
Wind speeds	0.3 to 1.0 m/s (~1 m height above mean surface). Increasing to about 130 m/s at 70 km altitude at 45° lat, somewhat slower at equator and decreasing to zero at the poles
Mean molecular weight*	43.45
Atmospheric composition	(near surface, by volume)
Major constituents (% by volume)*	96.5% carbon dioxide, 3.5% nitrogen
Minor constituents (ppm)	sulphur dioxide (150); argon (70) water (20); carbon monoxide (17); helium (12); neon (7)
Many other trace chemical species present	

* The bulk composition has been found to vary with altitude below 60 km with amount of nitrogen decreasing from 5% at 60 km to 3.4% at 22 km

APPENDIX II:

SUCCESSFUL SPACECRAFT MISSIONS TO VENUS

Mission	Agency	Launch date	Arrival/last contact	**Remarks**
Mariner 2	NASA	27 August 1962	Flyby: 14 December 1962–3 January 1963	First ever successful planetary mission. The infrared radiometer data showed limb darkening proving high surface temperatures.
Venera 4 entry probe	USSR	12 June 1967	18 October 1967	First entry into another planet's atmosphere by an instrumented spacecraft. Gave first indications of a thick carbon dioxide dominant atmosphere.
Mariner 5	NASA	14 June 1967	Flyby: 19 October 1967 Last contact: 5 November 1968	First demonstration of radio occultation technique to profile the ionosphere (electron density) and neutral atmosphere of a planet from the bending of the radio signal.
Venera 5 entry probe	USSR	5 January 1969	Entry: 16 May 1969	Second probe to enter atmosphere on the night side and send data from deep atmosphere (26.1 bar, about 25 km altitude) over its descent for 53 minutes.

Venera 6 entry probe	USSR	10 January 1969	Entry: 17 May 1969	Identical to Venera 5. Data transmitted for 51 minutes, down to about 10 km above the surface, also on the night side.
Venera 7 entry probe	USSR	17 August 1970	Landed: 15 December 1970	First probe to send data back from the surface of another planet.
Venera 8 entry probe	USSR	27 March 1972	Entry: 22 July 1972	First extensive measurements of the atmosphere during descent by a probe down to the surface.
Mariner 10	NASA	3 November 1973	Flyby: 5 February 1974	First spacecraft images of Venus confirming atmospheric superrotation and vortex organization of the global circulation over each pole.
Venera 9 lander and orbiter	USSR	8 June 1975	Lander entry: 22 October 1975	First images of the surface taken by a lander on the surface. First orbiter around Venus. Took some UV images with a scanning camera and first radio occultation investigation of the atmosphere by a Soviet orbiter. Also detected lightning (electrical activity) and thunder during probe descent.
Venera 10 lander and orbiter	USSR	14 June 1975	Lander entry: 25 October 1975	Identical to Venera 9, also sent pictures of the surface from another location.
Venera 11 lander	USSR	9 September 1978	Lander: 25 December 1978	Sampled the atmosphere during descent and from the surface for 95 minutes. Detected lightning (electrical activity) and thunder.

Venera 12 lander	USSR	14 September 1978	21 December 1978	Identical to Venera 11, sampled soil.
Pioneer Venus Multi-Probe	NASA	8 August 1978	Entry: 9 December 1978	A bus carried one larger probe and three small probes that returned data on the atmosphere from ~62 km altitude down to the surface.
Pioneer Venus orbiter	NASA	20 May 1978	Orbit insertion: 4 December 1978 Mission end: 8 October 1992	First radar to map a planet, extensive probing of the atmosphere by radio occultations and global imaging and polarimetry using spin-scan technique.
Venera 13 lander	USSR	30 October 1981	Lander: 1 March 1982	First colour panoramas from the surface; mass spectrometer
Venera 14 lander	USSR	4 November 1981	Lander: 5 March 1982	Identical to Venera 13, also sent colour images of the surface; taken after landing.
Venera 15 orbiter	USSR	2 June 1983	10 October 1983	Side-looking radar, first to use interferometric mapping.
Venera 16 orbiter	USSR	7 June 1983	11 October 1983	Identical to Venera 15, mapped northern hemisphere of Venus at 1–2 km resolution.

VeGa 1 balloon and lander	USSR	15 December 1984	Lander: 11 June 1985 Balloon sent data for 48 hours	First balloon floating in another planet's atmosphere. Collected temperature data and pressure data at about 54 km altitude for 48 hours and tracked by VLBI by a global network of twenty stations. Both landers detected presence of UV absorbers (220–400 nm) being present down to 47 km altitude. An X-ray fluorescence spectrometer also detected the presence of iron, sulphur and aluminium compounds.
VeGa 2 balloon and lander	USSR	21 December 1984	Lander: 15 June 1985 Balloon sent data for 48 hours	
Magellan orbiter	NASA	4 May 1989	Orbit insertion: 10 August 1990	Rising out of the cancelled Venus orbiter imaging radar, Magellan was built from spare parts and was launched from the Space Shuttle. Mapped 97 per cent of Venus's surface with ~200 m resolution and measured topography with ~100 m vertical resolution on 100 km footprint. First spacecraft to use aerobreaking to reduce the size of the orbit.
Galileo orbiter flyby	NASA	18 October 1989	Closest approach: 9 February 1990	First spacecraft images of Venus night side in near-infrared spectrum. Detection of electrical activity interpreted as due to lightning.

Cassini orbiter flyby	NASA/ ESA	15 October 1997	1st flyby: 26 April 1998 Second flyby: 24 June 1999	No detection of electrical activity related to lightning. The instrument was an improved version of the one flown on Galileo. EUV (80–120 nm) dayglow detected.
Venus Express orbiter	ESA	9 November 2005	Orbit insertion: 11 April 2006 Last contact: 19 January 2015	Extensive night side imaging of southern hemisphere in near-infrared. First solar and ultraviolet stellar occultation data on the atmospheric structure of Venus.
Akatsuki orbiter	JAXA	20 May 2010	First attempt: 7 December 2010 Second attempt: 7 December 2015 Still operating	Only spacecraft to get a second chance to enter orbit (JAXA's Nozomi failed to enter orbit on the second attempt after the first attempt failed). New morphologies discovered on the night side in near infrared data. First global images of the planet in thermal infrared. Possible optical detection of lightning.
Parker solar probe	NASA/ ESA	12 August 2018	1: 3 October 2018 2: 26 December 2019 3: 11 July 2020 4: 20 February 2021 5: 16 October 2021 6: 21 August 2023 7: 6 November 2024	Seven flybys of Venus over six years. Radio occultations in Ka band planned for later flybys.

BepiColombo	ESA/JAXA	20 October 2018	1: October 2020 2: August 2021	Collected Venus observations during first flyby. More observations planned for second flyby.
Solar orbiter	ESA/ NASA	10 February 2020	1: 26 December 2020 2: 8 August 2021 3: 3 September 2022 4: 18 February 2025 5: 24 December 2026	Five flybys of Venus. Some imaging of Venus and plasma environment observations will be conducted.
JUICE	ESA	December 2023	TBD	TBD

Appendix III:
Transits of Venus, 1631 to 2255

7 December 1631
4 December 1639
6 June 1761
3 June 1769
9 December 1874
6 December 1882
8 June 2004
6 June 2012
11 December 2117
8 December 2125
11 June 2247
9 June 2255

Note on Style, Constants, Units of Measurement and Abbreviations

The style conventions proposed by NASA and the Smithsonian for names of robotic spacecraft have been adopted throughout these chapters: italics are used only in cases of personal names for spacecraft, such as *Challenger*. For measurements mentioned within this book, we have used units from the metric system only, except for the inclusion of Fahrenheit conversions of Celsius.

The following is a list of data to clarify any information unfamiliar to readers.

Speed of light: 299,792,458 m/s

Time

1 second (s): 9,192,631,770 periods of the radiation involved in the unperturbed transition between the two hyperfine levels of the ground state of the caesium-133 atom

Terrestrial day:
mean sidereal (equinox to equinox) = 86,164.0907 s
mean rotation (fixed star to fixed star) = 86,164.0991 s
day (d) = 86,400 s
mean solar day = 86,400.002 s

Terrestrial year (yr): sidereal (fixed star to fixed star) = 365.256363 d
= 3.15 x 10^7 s

1 **megayear (Myr):** 10^6 yr
1 **gigayear (Gyr):** 10^3 Myr = 10^9 yr

Length

1 **astronomical unit (AU):** 1.495978707 x 10^{11} m = 9.314221863 x 10^7 miles

1 **metre (m):** the distance travelled by light in a vacuum in (299 782 458)$^{-1}$ s

1 **kilometre (km):** 10^3 m = 0.621371 miles

1 **mile★ (mi.):** 1.609344 km

1 **centimetre (cm):** 10^{-1} m = 0.393701 inches

1 **millimetre (mm):** 10^{-3} m

1 **micron (μm):** 10^{-6} m

1 **nanometre (nm):** 10^{-9} m

1 **angstrom★ (Å):** 0.1 nm

★indicates deprecated unit; unit on the right is preferred

Temperature

Kelvin scale is defined in terms of absolute zero where 1° Kelvin represents the same temperature difference as 1° on the Celsius scale. (Note that because it is an absolute scale, the Kelvin is not referred to or written as a degree.)

Celsius scale is defined in terms of absolute zero and the triple point of water (the pressure/temperature conditions where solid, liquid and vapour all coexist), such that absolute zero = 0 Kelvin = 273.15°C and the triple point of water is defined as 273.16 K.

Fahrenheit scale is a now deprecated system in which the freezing point of water is 32°F and the boiling point is 212°. (It is still the official scale used in the United States, its incorporated territories and a few other countries.) The conversion from Fahrenheit to

Centigrade is given by the formula C = (*f* – 32)°F × 5/9. From
Centigrade to Fahrenheit the formula is F = (*c* × 9/5) + 32°F.

Pressure

1 pascal (Pa): 1 newton/sq. m
1 bar: 100,000 Pa or slightly less than the current average
atmospheric pressure of Earth (~ 1.013 bar) = 14.5038 lb/sq. in.
1 millibar: 10^{-3} bar = 100 Pa

Angular measurements

2π radians = 360°
1° (degree) = 60′ (arcminutes)
1′ (arcminute) = 60″ (arcseconds)

atm = atmospheres

REFERENCES

1 THE 'EVENING STAR'

1 That Keats had Venus in mind in writing this sonnet rather than some other
object such as the North Star can be argued on internal evidence; only a
planet is 'steadfast' rather than twinkling, and when Venus is near its greatest
elongations it appears 'in lone splendour', far outshining every other planet
or star.
2 Otto Neugebauer, *The Exact Sciences in Antiquity* (New York, 1969), p. 81.
3 J.L.E. Dreyer, *A History of Astronomy from Thales to Kepler* [1906] (New York,
1953), pp. 4–5.
4 Jean Meeus, *More Mathematical Astronomy Morsels* (Richmond, VA, 2002),
p. 349.
5 Hans J. Nissen and Peter Heine, *From Mesopotamia to Iraq: A Concise History*
(Chicago, IL, and London, 2009), p. 23.
6 Patricia Monaghan, *Encyclopedia of Goddesses and Heroines* (Novato, CA, 2014), p. 39.
7 Nissen and Heine, *Mesopotamia to Iraq*, p. 58.
8 Roberta Binkley, 'Reading the Ancient Figure of Enheduanna', in *Rhetoric
Before and Beyond the Greeks*, ed. Carol S. Lipson and Roberta A. Binkley
(Albany, NY, 2004), p. 47.
9 See 'The Exaltation of Inanna: Opening Lines and Excerpts', www.
thehypertexts.com, accessed 17 August 2021.
10 John Ruskin, *Modern Painters* (London, 1906), vol. I, pp. 216–17.
11 Antonie Pannekoek, *A History of Astronomy* [1961] (New York, 1989), p. 33.
12 Joseph Campbell, *The Hero with a Thousand Faces*, 2nd edn (Princeton, NJ,
1968), p. 108n.
13 In John Milton, *Complete Poems and Major Prose*, ed. Merritt Y. Hughes
(Indianapolis, IN, and New York, 1957), book I, lines 437–40.
14 *Iliad*, xxii. In this passage, Athene has tricked Hector, who has been fleeing
Achilles (approaching, and glaring like the rising Sun), to stand his ground.

231

As Achilles charges forward, the glint of his spear appears like the evening star.

15 Translation by Mary Barnard, ed. and trans., *Sappho: A New Translation* (Oakland, CA, 2019), p. 14.

16 Alfred, Lord Tennyson, *In Memoriam A.H.H.*, CXXI; see www. poetryfoundation.org, accessed 17 August 2021.

17 According to the third-century CE biographer Diogenes Laertius (viii, 14), 'they say he (that is, Pythagoras) first said that Hesperos and Phosphoros are the same, as Parmenides says.'

18 Aristotle, *Metaphysics*, 986b.

19 Apart from the Pythagorean theorem, one of his other discoveries, and perhaps the most important, was that when the string of a musical instrument such as a cither or lyre is plucked, it produces a set of harmonics.

2 THE TELESCOPE: A NEW PHASE BEGINS

1 For full details, see William Sheehan, *Mercury* (London, 2018), pp. 13–28.

2 See Owen Gingerich, 'Galileo and the Phases of Venus', in *The Great Copernican Chase and Other Adventures in Astronomical History* (Cambridge, MA, and Cambridge, 1992), pp. 98–104 (p. 98).

3 Antonie Pannekoek, *A History of Astronomy* [1961] (New York, 1989), p. 35.

4 Ibid.

5 Ibid.

6 Patrick Moore, *The Planet Venus*, 3rd edn (London, 1961), p. 36.

7 At night the pupil diameter is greater, around 9 mm, and the theoretical resolution of the eye would be less. However, under those circumstances, the observation would be even more difficult, because of the overwhelming glare of the planet.

8 Moore, *The Planet Venus*, p. 36. Stephen J. O'Meara, personal communication, August 1988.

9 Sheehan, *Mercury*, p. 20.

10 William Sheehan and John Westfall, *Transits of Venus* (Amherst, NY, 2004), p. 67.

11 John Westfall and William Sheehan, *Celestial Shadows: Eclipses, Transits, and Occultations* (New York, 2015), p. 268.

12 Quoted in Sidney Bertram Gaythorpe, *Jeremiah Horrocks: 'The Pride and Boast of British Astronomy'*, ed. David Sellers (Leeds, 2020), p. 10.

13 Sheehan and Westfall, *Transits of Venus*, p. 82.

14 Ibid., p. 29.

15 Ibid., p. 84.

16 Ibid., p. 85.

17 Gaythorpe, *Jeremiah Horrocks*, p. 37.

3 ADVENTURES IN THE SEVENTEENTH AND EIGHTEENTH CENTURIES

1 F. Arago, *Astronomie Populaire* (Paris, 1827), vol. II, p. 533.

2 Sir John Frederick William Herschel, *Outlines of Astronomy*, 4th edn (Philadelphia, PA, 1861), p. 272.

3 William F. Denning, *Telescopic Work for Starlight Evenings* (London, 1891), pp. 146–7.

4 The definitive account is Helge Kragh, *The Moon That Wasn't: The Saga of Venus' Spurious Satellite* (Basel, 2008).

5 Christiaan Huygens, *Cosmotheoros, The Celestial Worlds Discover'd* (London, 1698), pp. 109–10.

6 Richard Baum and William Sheehan, 'G. D. Cassini and the Rotation of Venus', *Journal of the History of Astronomy*, XXIII (1992), pp. 299–301.

7 Francesco Bianchini, *New Phenomena of Hesperus and Phosphorus, or rather Observations Concerning the Planet Venus*, trans. S. Beaumont and P. Fay (London, 1996), p. 24.

8 An 'aerial' telescope was a type commonly used in the pre-achromatic era. First introduced about 1675 and employed by Christiaan Huygens and his brother Constantijn (it is not clear that they invented it), the lens was set on a mast or other support and let down, rotated, or extended to the eyepiece by means of a thread.

9 This can be inferred from modern spectrophotometric transmittance curves for the lenses made by Campani and other pioneering telescope-makers. See Giorgio Strano, *Galileo's Telescope: The Instrument That Changed the World* (Florence, 2008), p. 74. Venusian is the form most often used, though some authors still prefer the 'more tasteful' Cytherean – from Cythera, in Cyprus, near where the Greek goddess Aphrodite was supposed to have been born in the sea. Other terms used in the older literature include venusuvian, venerean and even occasionally venereal. We propose to stick with venusian.

10 Rev. T. W. Webb, *Celestial Objects for Common Telescopes* [1917], 6th edn (New York, 1962), vol. I, p. 68. One of us (W. S.) has seen such a globe, of uncertain provenance, in Christ Church library at Oxford.

11 See Sheehan, *Mercury*, p. 24.

12 See Sheehan and Westfall, *Transits of Venus*, Chapters Eight and Nine.

13 Quoted ibid., p. 152.

14 G. Schneider, J. M. Pasachoff and Leon Golub, 'TRACE Observations of the 15 November 1999 Transit of Mercury and the Black Drop Effect: Considerations for the 2004 Transit of Venus', *Icarus*, CLXIII (2004), pp. 249–56.

15 The claim was vigorously promoted by the famous Soviet astronomer V. V. Sharanov. See 'Lomonosov as an Organizer of the Transit of Venus in 1761 in

Russia and his Discovery of the Atmosphere of Venus', in M. V. Lomonosov, *Complete Works*, ed. S. Vavilov and T. Kravets (Moscow and Leningrad, 1955), vol. IV, pp. 7–40.

16 Jay M. Pasachoff and William Sheehan, 'Lomonosov, The Discovery of Venus's Atmosphere, and Eighteenth-Century Transits of Venus', *Journal of Astronomical History and Heritage*, XV (2012), pp. 3–14.

17 P. Tanga et al., 'Sunlight Refraction in the Mesosphere of Venus during the Transit on June 8th, 2004', *Icarus*, CCXVIII (2012), pp. 207–19.

4 A Planet of Illusions – and a Few Facts

1 E. E. Barnard, 'Physical and Micrometrical Observations of the Planet Venus Made at the Lick Observatory with the 12-inch and 36-inch Refractors', *Astrophysical Journal*, V (1897), p. 300.

2 Often Schröter in the English-speaking literature, though Schroeter is to be preferred, since this is the form the astronomer himself always used.

3 Richard Baum, *The Haunted Observatory: Curiosities from the Astronomer's Cabinet* (Amherst, NY, 2007), p. 147.

4 H. N. Russell's method of using the observed extensions to estimate the atmospheric height and density of the venusian atmosphere was published in his paper, 'The Atmosphere of Venus', *Astrophysical Journal*, IX (1899), pp. 284–99.

5 Baum, *Haunted Observatory*, p. 162.

6 Ibid., p. 137.

7 Willy Ley, *Rockets, Missiles and Space Travel* (New York, 1951), p. 36.

8 Richard Baum, 'Franz von Paula Gruithuisen and the Discovery of the Polar Spots of Venus', *Journal of the British Astronomical Association*, CV (1995), pp. 144–7.

9 E. M. Antoniadi, 'Notes on the Rotation Period of Venus', *Monthly Notices of the Royal Astronomical Society*, LVIII (1898), p. 313.

10 Barnard, 'Physical and Micrometrical Observations', p. 299.

11 See Sheehan, *Mercury*, pp. 41–6.

12 P. Lowell, 'The Rotation Period of Venus', *Astronomische Nachrichten*, 3406 (1897), col. 361.

13 E. M. Antoniadi, *English Mechanic*, VI (1898), p. 474.

14 William Graves Hoyt, *Lowell and Mars* (Tucson, AZ, 1976), pp. 110–11.

15 William Sheehan and Thomas Dobbins, 'The Spokes of Venus: An Illusion Explained', *Journal for the History of Astronomy*, XXXIV (2003), pp. 53–64.

16 P. Lowell, *Mars and Its Canals* (New York, 1906), p. 178n.

17 J. M. [Mc] Harg, *Gazette Astronomique*, I (1908).

18 Frank E. Seagrave, 'Venus', *Popular Astronomy*, XXVII (1919), p. 406.

19 W. W. Coblentz, *From the Life of a Researcher* (New York, 1951), pp. 158–9.

20 Percival Lowell, *The Evolution of Worlds* (New York, 1909), p. 80.

21 For a list, see Patrick Moore, *The Planet Venus*, 3rd edn (London, 1961), Appendix 2, 'Estimated Rotation Periods', p. 133.

22 F. E. Ross, 'Photographs of Venus', *Astrophysical Journal*, VI (1928), p. 82.

23 Audouin Dollfus, 'Visual and Photographic Studies of Planets at the Pic du Midi', in *Planets and Satellites*, ed. G. P. Kuiper and B. M. Middlehurst (Chicago, IL, 1961), p. 554.

24 For this and the following section, see William Sheehan and Thomas Dobbins, 'Charles Boyer and the Clouds of Venus', *Sky and Telescope*, XCVII (1999), pp. 56–60.

25 Ibid., p. 60.

26 Ibid.

5 CHASING SPECTRA

1 See Stephen Case, *Making Stars Physical: The Astronomy of Sir John Herschel* (Pittsburgh, PA, 2018).

2 Jules Janssen, 'Sur l'absorption de la chaleur rayonnante obscure dans les milieu de l'oeil', *Annales de Physique et de Chimie*, 3rd series, LX (1860), pp. 71–93.

3 Françoise Launay, *The Astronomer Jules Janssen: A Globetrotter of Celestial Physics* (New York, 2012), p. 31.

4 For more details, see William Sheehan and John Westfall, *Transits of Venus* (Amherst, NY, 2004).

5 Ibid., p. 115.

6 See 'Pierre-Jules-César Janssen', *Who's Who of Victorian Cinema*, www. victorian-cinema.net/janssen, accessed 18 August 2021.

7 Sheehan and Westfall, *Transits of Venus*, p. 255.

8 A comprehensive account of the nineteenth-century British transit expeditions is Jessica Ratcliff, *The Transit of Venus Enterprise in Victorian Britain* (London, 2008).

9 Sheehan and Westfall, *Transits of Venus*, p. 278.

10 Launay, *Jules Janssen*, p. 118.

11 Ibid., p. 119.

12 W. S. Adams and T. Dunham, 'Absorption Bands in the Infra-Red Spectrum of Venus', *Publications of the Astronomical Society of the Pacific*, XLIV (1932), pp. 243–5.

13 Arthur Adel, American Institute of Physics Oral History Interview with Robert W. Smith, 12 August 1987, www.aip.org, accessed 10 August 2021.

14 Audouin Dollfus, 'Pioneering Balloon Astronomy in France', *Sky and Telescope*, LXVI (1983), pp. 381–6.

15 J. Strong, J., M. D. Ross and C. B. Moore, 'Some Observations of the Atmosphere of Venus and the Earth during the Strato Lab IV Balloon Flight', *Journal of Geophysical Research*, LXV (1960), p. 2526.

6 A New Era: Radar and Spacecraft

1 Carl Sagan, 'The Radiation Balance of Venus', Technical Report no. 32–4 (1960).
2 Andrew Freedman and Chris Mooney, 'Earth's Carbon Dioxide Levels Hit Record High, Despite Coronavirus-Related Emissions Drop', www.washingtonpost.com, 4 June 2020.
3 R. M. Goldstein and R. L. Carpenter, 'Rotation of Venus: Period Estimated from Radar Measurements', *Science*, CXXXIX/3558 (1963), pp. 910–11.
4 I. I. Shapiro, 'Resonance Rotation of Venus', *Science*, CLVII (1967), pp. 423–5.
5 It was common in the early days of planetary exploration to build two identical spacecraft so that in case of failure, the second could be used. The U.S. missions did not launch the second unit, but the Soviets tended to launch both, resulting in two spacecraft arriving just days apart: Venera 7 and 8, Venera 9 and 10, Venera 13 and 14, Venera 15 and 16, as well as the final missions, VeGa 1 and VeGa 2.
6 Carl Sagan, 'The Radiation Balance of Venus', Technical Report no. 32–4 (15 September 1960).
7 As noted by Robert Reeves, *The Superpower Space Race: An Explosive Rivalry through the Solar System* (New York, 1994), p. 170.
8 V. S. Avduevsky et al., 'Model of the Atmosphere of the Planet Venus based on Results of Measurements Made by the Soviet Interplanetary Station Venera 4', *Journal of the Atmospheric Sciences*, XXV (1968), pp. 537–45.
9 V. S. Avduevsky, M. Ya. Marov and M. K. Rozhdestvensky, 'A Tentative Model of the Venus Atmosphere Based on the Measurements of Venera 5 and 6', *Journal of the Atmospheric Sciences*, XXVII (1970), pp. 561–8.
10 Peter D. Ward, *Under a Green Sky* (Washington, DC, 2007), p. xii.
11 See J. Touma and J. Wisdom, 'Evolution of the Earth–Moon System', *Astronomical Journal*, CVIII (1994), pp. 1943–61.
12 Alexandre C. M. Correia and Jacques Laskar, 'The Four Final Rotation States of Venus', *Nature*, CDXI/6839L (2001), pp. 767–70.
13 Bruce Bills, 'Variations in the Rotation Rate of Venus Due to Orbital Eccentricity Modulation of Solar Torques', *Journal of Geophysical Research*, CX:E11007 (2005).

7 What Are the Clouds Made Of?

1 The Venera probes (5–8) had already shown the superrotation of the deep atmosphere from line-of-sight Doppler tracking. The four probes on the Pioneer Venus Multiprobe mission in 1978 used Differential Long Baseline Interferometry (DLBI) tracking that did not rely on line-of-sight Doppler and allowed the north–south component to be determined by all four probes. However, these were below 64 km. Mariner 10 provided the latitudinal profiles of the east–west and north–south winds at 70 km, the level of the cloud tops.

2 J. E. Ainsworth and J. R. Herman, 'Venus Wind and Temperature Structure: The Venera 8 Data', *Journal of Geophysical Research*, LXXX (1975), pp. 173–9.

3 The retrograde and prograde terminology can be misused. To make things clear, the terms are defined according to whether the spin vector due to rotation is aligned with the orbital revolution vector or not. Since the Venus winds in the bulk of the atmosphere blow in the same direction as the solid planet, some believe that they should be called prograde not retrograde. Of course this leads to confusing descriptions of the wind direction in the thermosphere. Unfortunately some ambiguous terminology was used to describe the sub-solar to antisolar flow found in that part with overlaid superrotation.

4 V. E. Suomi and S. S. Limaye, 'Venus – Further Evidence of Vortex Circulation', *Science*, CCI (1978), pp. 1009–11.

5 S. S. Limaye et al., 'Vortex Circulation on Venus: Dynamical Similarities with Terrestrial Hurricanes', *Geophysical Research Letters*, XXXVI (2009), L04204.

6 M.J.S. Belton et al., 'Images from Galileo of the Venus Cloud Deck', *Science*, New Series, CCLIII/5027 (1991), pp. 1531–6.

7 D. A. Allen and J. W. Crawford, 'Cloud Structure on the Dark Side of Venus', *Nature*, CCCVII (1984), pp. 222–4.

8 D. Crisp et al., 'The Dark Side of Venus: Near-Infrared Images and Spectra from the Anglo-Australian Observatory', *Science*, CCLIII (1991), pp. 1263–6.

9 J. Lecacheux et al., 'Detection of the Surface of Venus at 1.0 μm from Ground-Based Observations', *Planetary and Space Science*, XLI (1993), pp. 543–9.

10 S. E. Smrekar et al., 'Recent Hotspot Volcanism on Venus from VIRTIS Emissivity Data', *Science*, CCCXXVIII/5978 (2010), p. 605; E. V. Shalygin, W. G. Markiewicz et al., 'Active Volcanism on Venus in the Ganiki Chasma Rift Zone', *Geophysical Research Letters*, XLII (2015), pp. 4762–9; N. T. Mueller et al., 'Search for Active Lava Flows with VIRTIS on Venus Express', *Journal of Geophysical Research (Planets)*, CXXII (2017), p. 1021.

11 V. Wilquet et al., 'Optical Extinction due to Aerosols in the Upper Haze of Venus: Four Years of SOIR/VEX Observations from 2006 to 2010', *Icarus*,

CCXVII (2012), pp. 875–81; S. Limaye et al., 'Focal Lengths of Venus Monitoring Camera from Limb Locations', *Planetary and Space Science*, CXIII (2015), pp. 169–83.

12 M. Shimizu, 'Ultraviolet Absorbers in the Venus Clouds', *Astrophysics and Space Science*, 51 (1977), p. 497.

13 L. G. Young et al., 'The Planet Venus: A New Periodic Spectrum Variable', *Astrophysical Journal*, CLXXXI (1973), L5.

8 THE SURFACE OF VENUS

1 Lianne Kolirin, 'Venus Is a Russian Planet – Say the Russians', CNN, 18 September 2020.

2 A Venera 14 recording of wind sounds at Venus's surface can be heard at '40-year-old Sound Recording from Venus Planet of Venera Mission', www.youtube.com, accessed 2 November 2020.

3 M. A. Kolosov et al., 'Characteristics of the Surface and Features of the Propagation of Radio Waves in the Atmosphere of Venus from Data of Bistatic Radiolocation Experiments Using Venera 9 and 10 Satellites', *Icarus*, XLVIII (1981), pp. 188–200.

4 G. G. Schaber et al., 'Geology and Distribution of Impact Craters on Venus: What Are They Telling Us?', *Journal of Geophysical Research*, XCVII (1992), pp. 13257–302.

5 W. B. McKinnon et al., 'Cratering on Venus: Models and Observations', in *Venus II: Geology, Geophysics, Atmosphere, and Solar Wind Environment*, ed. S. W. Bougher, D. M. Hunten and R. J. Phillips (Tucson, AZ, 1997), pp. 969–1014.

6 M. A. Ivanov and J. W. Head, 'Global Geological Map of Venus', *Planetary and Space Science*, LIX (2011), pp. 1559–600; M. A. Ivanov and J. W. Head, 'The History of Tectonism on Venus: A Stratigraphic Analysis', *Planetary and Space Science*, CXIII–CXIV (2015), pp. 10–32; M. A. Ivanov and J. W. Head, 'The History of Volcanism on Venus', *Planetary and Space Science*, LXXXIV (2013), pp. 66–92.

7 R. G. Strom, G. G. Schaber and D. D. Dawson, 'The Global Resurfacing of Venus', *Journal of Geophysical Research*, XCIX (E5) (1994), pp. 10899–926.

8 A. T. Basilevsky and J. W. Head, 'The Surface of Venus', *Reports on Progress in Physics*, LXVI (2003), pp. 1699–1734.

9 J. W. Head et al., 'Venus Volcanism: Classification of Volcanic Features and Structures, Associations, and Global Distribution from Magellan Data', *Journal of Geophysical Research*, XCVII (1992), pp. 13153–98.

10 E. V. Shalygin et al., 'Active Volcanism on Venus in the Ganis Chasma Rift Zone', *Geophysical Research Letters*, XLII (2015), pp. 4762–9.

11 E. Grosfils et al., 'Geologic Map of the Ganiki Planitia Quadrangle (V-14), Venus', Scientific Investigations Map 3121, U.S. Geological Survey

(Washington, DC, 2011); E. M. Venechuk et al., 'Analysis of Tectonic Lineaments in the Ganiki Planitia (V-14) Quadrangle, Venus', *Lunar and Planetary Science*, XXXVI (2005).

12 J. B. Garvin et al., 'DAVINCI+: Deep Atmosphere of Venus Investigation of Noble Gases, Chemistry, and Imaging Plus, Lunar and Planetary Science Conference' (2020), https://ui.adsabs.harvard.edu.

13 R. Ghail et al., 'The Science Goals of the EnVision Venus Orbiter Mission' (2020), https://ui.adsabs.harvard.edu.

14 S. Smrekar et al., 'VERITAS (Venus Emissivity, Radio Science, InSAR, Topography and Spectroscopy): A Proposed Discovery Mission' (2020), https://ui.adsabs.harvard.edu.

15 McKinnon et al., 'Cratering on Venus', p. 981.

16 M. J. Way et al., 'Was Venus the First Habitable World of Our Solar System?', *Geophysical Research Letters*, XLIII (2016), pp. 8376–83.

17 B. M. Jakosky et al., 'Mars' Atmospheric History Derived from Upper-Atmosphere Measurements', *Science*, CCCLV (2017), pp. 1408–10.

18 R. A. Craddock and A. D. Howard, 'The Case for Rainfall on a Warm, Wet Early Mars', *Journal of Geophysical Research: Planets* (2002), https://doi.org.

9 LIFE ON VENUS?

1 For the Arrhenius quote, see Svante Arrhenius, *The Destinies of the Stars*, trans. Joens Elias Fries (New York, 1918), pp. 316–20. For the spectrum of Venus, see L. G. Young, 'Infrared Spectra of Venus', in *Exploration of the Planetary System*, Proceedings of the Symposium, Torun, Poland, 5–8 September 1973 (Dordrecht, 1974), pp. 77–160.

2 Fred Hoyle, *Frontiers of Astronomy* (New York, 1955), p. 71.

3 W. F. Libby and P. Corneil, 'Water on Venus?', in *Planetary Atmospheres*, ed. C. Sagan, T. C. Owen and H. J. Smith, Proceedings from 40th IAU Symposium, Marfa, Texas, 26–31 October 1969 (Dordrecht, 1971), pp. 51–61.

4 J. Seckbach and W. F. Libby, 'Vegetative Life on Venus? Or Investigations with Algae Which Grow under Pure CO_2 in Hot Acid Media and at Elevated Pressures', in *Planetary Atmospheres*, ed. Sagan, Owen and Smith, pp. 62–83.

5 H. Morowitz and C. Sagan, 'Life on the Surface of Venus?', *Nature*, CCXVI (1967), pp. 1198–9.

6 W. B. McKinnon et al., 'Cratering on Venus: Models and Observations', in *Venus II: Geology, Geophysics, Atmosphere, and Solar Wind Environment*, ed. S. W. Bougher, D. M. Hunten and R. J. Phillips (Tucson, AZ, 1997), pp. 969–1014.

7 M. A. Ivanov and J. W. Head, 'Global Geological Map of Venus', *Planetary and Space Science*, LIX (2011), pp. 1559–600; M. A. Ivanov and J. W. Head, 'The History of Tectonism on Venus: A Stratigraphic Analysis', *Planetary and Space*

Science, CXIII–CXIV (2015), pp. 10–32; M. A. Ivanov and J. W. Head, 'The History of Volcanism on Venus', *Planetary and Space Science*, LXXIV (2013), pp. 66–92.

8 T. M. Donahue et al., 'Venus Was Wet: A Measurement of the Ratio of Deuterium to Hydrogen', *Science*, CCXVI (1982), pp. 630–33; T. M. Donahue and R. R. Hodges Jr, 'Past and Present Water Budget of Venus', *Journal of Geophysical Research*, XCVII (1992), pp. 6083–91.

9 A. Fedorova et al., 'HDO and H2O Vertical Distributions and Isotopic Ratio in the Venus Mesosphere by Solar Occultation at Infrared Spectrometer on Board Venus Express', *Journal of Geophysical Research (Planets)*, CXIII (2008), pp. E00B22.ff10.1029/2008JE003146.

10 M. Persson et al., 'H+/O+ Escape Rate Ratio in the Venus Magnetotail and Its Dependence on the Solar Cycle', *Geophysical Research Letters*, XLV (2018), pp. 10,805–11.

11 J. F. Kasting, 'Runaway and Moist Greenhouse Atmospheres and the Evolution of Earth and Venus', *Icarus*, LXXIV (1988), pp. 472–94.

12 M. J. Way et al., 'Was Venus the First Habitable World of Our Solar System?', *Geophysical Research Letters*, XLIII (2016), pp. 8376–83.

13 B. M. Jakosky et al., 'Mars' Atmospheric History Derived from Upper-Atmosphere Measurements of 38Ar/36Ar' *Science*, CCCLV (2017), pp. 1408–10; A. A. Fedorova et al., 'Stormy Water on Mars: The Distribution and Saturation of Atmospheric Water during the Dusty Season', *Science*, CCCLXVII (2020), pp. 297–300.

14 Barbara Cavalazzi and Frances Westall, eds, *Biosignatures for Astrobiology* (New York, 2019).

15 B. Hapke and R. Nelson, 'Evidence for an Elemental Sulfur Component of the Clouds from Venus Spectrophotometry', *Journal of Atmospheric Sciences*, XXXII (1975), pp. 1212–18.

16 Shimizu, 'Ultraviolet Absorbers', p. 499.

17 C. Boyer, 'La haute atmosphère de Venus: tentative d'explication de sa rotation', *L'Astronomie*, 100 (1986), p. 77.

18 David Harry Grinspoon, *Venus Revealed: A New Look below the Clouds of Our Mysterious Twin Planet* (Reading, MA, 1997), p. 315.

19 Charles Darwin to Joseph Dalton Hooker, 1 February 1871, Cambridge University: Darwin Correspondence Project, www.darwinproject.ac.uk/letter/DCP-LETT-7471.xml.

20 I. Daniel, P. Oger and R. Winter, 'Origins of Life and Biochemistry under High-Pressure Conditions', *Chemical Society Reviews*, XXXV (2006), pp. 858–75.

21 Michael Marshall, 'How the First Life on Earth Survived Its Biggest Threat – Water', *Nature*, www.nature.com, 9 December 2020.

22 A good primer is Michael Gross, *Life on the Edge: Amazing Creatures Thriving in Extreme Environments* (New York and London, 1998).

23 R. G. Prinn and B. Fegley, 'The Atmospheres of Venus, Earth, and Mars: A Critical Comparison', *Annual Review of Earth and Planetary Sciences*, XV (1987), 171.10.1146/annurev.ea.15.050187.001131.

24 Based on data from L. R. Bakken and R. A. Olsen, 'Buoyant Densities and Dry-Matter Contents of Microorganisms: Conversion of a Measured Biovolume into Biomass', *Applied Environmental Microbiology*, XLV (1983), pp. 1188–95.

25 D. Grinspoon and M. Bullock, 'Astrobiology and Venus Exploration', in *Exploring Venus as a Terrestrial Planet*, ed. Larry W. Esposito, Ellen R. Stofan and Thomas E. Cravens (Washington, DC, 2007), pp. 191–206.

26 S. Seager et al., 'The Venusian Lower Atmosphere Haze as a Depot for Desiccated Microbial Life: A Proposed Life Cycle for Persistence of the Venusian Aerial Biosphere', *Astrobiology*, 10.1089/ast.2020.2244

27 D. Crisp, 'Radiative Forcing of the Venus Mesosphere: 1. Solar Fluxes and Heating Rates', *Icarus*, LXVII (1986), pp. 484–514.

28 S. S. Limaye et al., 'Venus' Spectral Signatures and the Potential for Life in the Clouds', *Astrobiology*, 18 (2018), pp. 1181–98.

29 S. Seager, M. Schrenk and W. Bains, 'An Astrophysical View of Earth-Based Metabolic Biosignature Gases', *Astrobiology*, XII (2012), pp. 61–82; C. Sousa-Silva et al, 'Phosphine as a Biosignature Gas in Exoplanet Atmospheres', *Astrobiology*, XX (2020), pp. 235–68.

30 T. Encrenaz et al., 'A Stringent Upper Limit of the PH_3 Abundance at the Cloud Top of Venus', *Astronomy and Astrophysics*, CDXLIII (2020), L5.10.1051/0004-6361/202039559; J. S. Greaves et al., 'Phosphine Gas in the Cloud Decks of Venus', *Nature Astronomy* (2020), 10.1038/s41550-020-1174-4; J. S. Greaves et al., 'Re-analysis of Phosphine in Venus' Clouds', arXiv:.org>astro-ph>arXiv:2011.08176.

31 Y. J. Lee et al., 'Long-Term Variations of Venus's 365 nm Albedo Observed by Venus Express, Akatsuki, MESSENGER, and the Hubble Space Telescope', *Astronomical Journal*, CLVIII/126 (September 2019).

10 Observing Venus

1 Quoted in William Sheehan, *The Immortal Fire Within: The Life and Work of Edward Emerson Barnard* (Cambridge, 1995), p. 16.

2 Patrick Moore, *The Planet Venus*, 3rd edn (New York, 1961), pp. 87–8.

3 Bradford A. Smith to Richard Baum, 15 February 1960; courtesy Richard Baum.

4 S. S. Limaye et al., 'Venus Looks Different from Day to Night across Wavelengths: Morphology from Akatsuki Multispectral Images', *Earth, Planets and Space*, LXX/24 (2018), https://doi.org/10.1186/s40623-018-0789-5.

5 F. Behar-Cohen et al., 'Ultraviolet Damage to the Eye Revisited', *Clinical Ophthalmology*, VIII (2014), pp. 87–104.

6 Ewen A. Whitaker, 'Visual Observations of Venus in the UV', *Journal of the British Astronomical Association*, XCI (1989), pp. 296–7.

7 Thomas A. Dobbins, Donald C. Parker and Charles F. Capen, *Introduction to Observing and Photographing the Solar System* (Richmond, VA, 1988), pp. 39–40.

8 D. A. Allen and J. W. Crawford, 'Cloud Structure on the Dark Side of Venus', *Nature*, CCCVII (1984), pp. 222–4.

9 For a recent review, see W. Sheehan, K. Brasch, D. Cruikshank and R. Baum, 'The Ashen Light: The Oldest Unsolved Solar System Mystery', *Journal of the British Astronomical Association*, CXXIV (2014), pp. 209–15. For an upcoming publication, see John Barentine, *Mystery of the Ashen Light of Venus: Investigating a 400-Year-Old Phenomenon* (Cham, 2021).

10 Richard Baum and David Graham, 'Mercury and Venus', in *The British Astronomical Association Observing Guide, 2002*, ed. Richard J. McKim (London, 2002), p. 14.

FURTHER READING AND RESOURCES

VENUS

Bengtsson, L., et al., eds, *Towards Understanding the Climate of Venus* (New York, Heidelberg, Dordrecht and London, 2013)

Bouger, S.W., D. M. Hunten and R. J. Phillips, eds, *Venus*, vol. II: *Geology, Geophysics, Atmosphere, and Solar Wind Environment* (Tucson, AZ, 1997)

Cattermole, Peter, and Patrick Moore, *Atlas of Venus* (Cambridge, 1997)

Grinspoon, David H., *Venus Revealed: A New Look below the Clouds of Our Mysterious Twin Planet* (Reading, MA, 1997)

Hunten, D. M., et al., *Venus* (Tucson, AZ, 1983)

Mackwell, S. J., et al., *Comparative Climatology of Terrestrial Planets* (Tucson, AZ, 2013)

Meadows, V. S., et al., *Planetary Astrobiology* (Tucson, AZ, and Houston, TX, 2020)

Marov, Mikhail Ya., et al., *The Planet Venus* (New Haven, CT, 1998)

Moore, Patrick, *The Planet Venus*, 1st edn (London, 1956)

Roth, Ladislav E., and Stephen D. Wall, *The Face of Venus: The Magellan Radar-Mapping Mission* (Washington, DC, 1995)

TRANSITS OF VENUS

Lomb, Nick, *Transit of Venus 1631 to the Present* (New York, 2011)

Sellers, David, *The Transit of Venus: The Quest to Find the True Distance of the Sun* (Leeds, 2001)

Sheehan, William, and John Westfall, *The Transits of Venus* (Amherst, NY, 2004)

Westfall, John, and William Sheehan, *Celestial Shadows: Eclipses, Transits, and Occultations* (New York, Heidelberg, Dordrecht and London, 2015)

MAPS

Venus Map Catalog, U.S. Geological Survey (listed and downloadable at www.lpi.usra.edu/resources/venus_maps)

Acknowledgements

William Sheehan

In addition to my family, I express particular thanks to my co-author Sanjay Limaye, for enlightening me about many a thorny problem of Venus research. Thanks to Peter Morris, series editor of the Kosmos series; Amy Salter, desk editor at Reaktion Books; and publisher Michael Leaman, all of whom were indispensable to making this book a reality. Others who made valuable contributions include Lauren Amundson of Lowell Observatory archives, Richard Baum, Klaus Brasch, Henri Camichel, Dale P. Cruikshank, Audouin Dollfus, Peter Hingley, Françoise Launay, Anthony Misch, Jay Pasachoff, Christophe Pellier, Paolo Tanga, Sebastien Voltmer, Sean Walker, Ewen A. Whitaker, Thomas Widemann and John A. Westfall.

Sanjay Limaye

I thank my family for tolerating my pursuit of mysteries that Venus presents us. We would not have learned about these mysteries without the efforts of unknown numbers of inventive engineers and scientists all over, but particularly in the Soviet Union for their pioneering contributions through the Venera and VeGa missions as well as the Mariner and Pioneer missions to Venus. The many whose contributions to understanding Venus have been personally influential include Verner Suomi, Jacques Blamont, Vassily Moroz, Leonid Ksanfomality, Tom Donahue, Hal Masursky, Arv Kliore, Carl Sagan, Conway Leovy and Al Seiff. Wojtek Markiewicz's selfless dedicated efforts with the Venus Monitoring Camera on Venus Express orbiter with Dima Titov provided multi-wavelength global image coverage of Venus over more than eight years which led to many insights about Venus clouds. Masato Nakamura and the Akatsuki team led by Takeshi Imamura and Takehiko Sato at ISAS provided many opportunities to dig deeper into the Venus mysteries.

It is a pleasure to acknowledge Jim Hansen, Larry Travis, Kiyoshi Kawabata, Makiko and Makoto Sato, Lex Lane and Andy Lacis, whom I had the opportunity to learn from at GISS while working with the Pioneer Venus orbiter data; Bob Krauss, who worked with the Mariner 10 project for Venus image data acquisition; and Larry Sromovsky and Hank Revercomb, who formed the Planetary Group with Vern Suomi at Space Science and Engineering Center. Over the years it has been a pleasure to discuss Venus with Lusia Zasova, Jim Head, Steve Saunders and Adriana Ocampo, and thanks to Jim Green for encouraging me with VEXAG and support for promoting Venus exploration internationally.

I thank Bill Sheehan for approaching me for this book, and I continue to discover that his grasp and breadth of knowledge about planets throughout history is far beyond what I had gathered over the decades since we first met. Writing this book with him has been a pleasurable learning experience. Finally, Rosalyn Perztborn's broad knowledge and support in Venus outreach to a wide audience have been invaluable.

Photo Acknowledgements

The author and publishers wish to thank the organizations and individuals listed below for authorizing reproduction of their work:

Advances in Space Research, VII/12 (1987): p. 157 (Venera 15 and Venera 16 radar data: O.vN. Rzhiga/Images and maps of Venus processed by Don P. Mitchell); Astronomische Nachrichten: pp. 67 top (1891, No. 3018, Column 295), 74 (1897); Astrophysical Journal: pp. 64 (1897), 82 (Plate 1, 1928); L'Astronomie: p. 16; Bianchini, Hesperi et phosphori nova phaenomena . . . (Rome, 1728): pp. 49, 50 left, 50 top right; Butrica: p. 122 top; D. B. Campbell et al., 'Radar Interferometric Observations of Venus at 70-Centimeter Wavelength', Science, 170 (1970), p. 1090: p. 122 centre right; Terence Dickinson: p. 10; Audouin Dollfus: pp. 112, 113; Audouin Dollfus, Observatoire de Paris: pp. 18, 66, 84, 87, 88 top and top centre, 89 (above: from Mariner 10; below: from images by C. Boyer and P. Guérin); Helge Drange: p. 120; ESA: pp. 145, 146 centre, 175, 178; Camille Flammarion, Les Ciel et terres (Paris, 1884): pp. 30, 31 top, 45 top right, 70, 71; Gazette Astronomique (1908): p. 76; Kevin M. Gill, NASA/JPL-Caltech: p. 143 left and right; Michael Hammer, University of Arizona: p. 169; Head et al., 1992: p. 174; ISAS/JAXA: p. 199 (Akatsuki UVI data, Planet-C/Akatsuki Project); JAXA: p. 177; JAXA/ISAS/DARTS/Damia Bouic: p. 148 top left; Emily Lakdawalla – NASA/JPL: pp. 164 top, 170; Françoise Launay: p. 101; William Leatherbarrow: p. 211; Lick Observatory, University of California: p. 105 centre left; Sanjay Limaye: pp. 140 top and bottom, 155; Sanjay Limaye and Rakesh Mogul: p. 200; Lowell Observatory Archives: pp. 45 top left, 75, 111; Patrick Moore, The Planet Venus (London): pp. 29 bottom (3rd edn, 1961), 48 (1st edn, 1956); NASA: pp. 127 bottom right, 146 top, 148 top right, 172, 173 bottom; NASA/GSFC: pp. 180, 181 (visualization by CI Labs, Michael Lentz and others); NASA/JPL: pp. 126, 127 top, 152, 159, 161, 164 bottom left, 165, 167, 171, 173 top, 181 bottom; NASA/JPL-Caltech: pp. 142, 163; New Mexico State University Library Archives and Special Collections: p. 208; New York Times Sunday Magazine, 24 March 1912: p. 187; Jay Pasachoff: pp. 41 bottom, 105 bottom right; Christophe Pellier:

p. 213; Henry Chamberlaine Russell, *Observations of the Transit of Venus, 9 December, 1874*: pp. 59 top left and top right, 102 (Sydney, NSW, Australia, Charles Potter, Government Printer, 1892); Schaber et al. (1992): p. 162; Johann Hieronymus Schroeter, *Selenotopographischen fragmenten über den Mond*: p. 68 (Band 1, 1791); William Sheehan: pp. 15, 21, 34, 51; William Sheehan collection: pp. 69, 103; David E. Soper, Institute of Theoretical Science, University of Oregon: p. 166; Ted Stryk: p. 154 centre and bottom (www.planetary.org/articles/0724-standing-on-venus-in-1975); Paolo Tanga of the Côte d'Azur Observatory, Nice, France: pp. 106, 107 top and bottom, 110; University of Wisconsin-Madison Archives: p. 135; U.S. Department of the Interior-U.S. Geological Survey, 2011: p. 176; Sebastian Voltmer: pp. 42, 67 centre, 88 bottom centre; Sean Walker: p. 31 bottom; Sean Walker, *Sky and Telescope*: p. 212; Michael Way, NASA Goddard Institute for Space Studies: p. 192; John E. Westfall: pp. 54, 57; Ewen A. Whitaker: p. 210; Yerkes Observatory: p. 81.

Jastrow, the copyright holder of the image on p. 20, has released it into the public domain. (Louvre Museum, Paris/Excavated by Jacques de Morgan.); Fæ, the copyright holder of the image on p. 22, has published it online under conditions imposed by a Creative Commons Attribution-Share Alike 3.0 Unported License. (British Museum.); Daniel V. Schroeder has licensed the image on p. 29 under a Creative Commons Attribution-Non Commercial 3.0 Unported License. (Attribution: Daniel V. Schroder/link to the original at https://physics.weber. edu/schroeder/ua/BeforeCopernicus.html); Sailko, the copyright holder of the image on p. 33, has published it online under conditions imposed by a Creative Commons Attribution 3.0 Unported License; Chuck Bueter, the copyright holder of the image on p. 41 top, has published it online under conditions imposed by a Creative Commons Attribution-Share Alike 3.0 Unported License; Pascalou petit, the copyright holder of the image on p. 83, has published it online under conditions imposed by Creative Commons Attribution-Share Alike 3.0 Unported, 2.5 Generic, 2.0 Generic and 1.0 Generic Licenses; Page 95 top: This is a *retouched picture*, which means that it has been digitally altered from its original version. The original can be viewed here: Fraunhofer lines DE.svg. Modifications made by Cepheiden. The copyright holder has released the work into the public domain; Spectrum of blue sky.png, the copyright holder of the image on p. 95 centre right, has published it online under conditions imposed by Creative Commons Attribution-Share Alike 4.0 International, 3.0 Unported, 2.5 Generic, 2.0 Generic and 1.0 Generic Licenses; Page 100: This work, dated 1874, by an unknown author is in the public domain; Page 117: Image taken from en.wikipedia en:Electromagnetic-Spectrum.svg and en:Electromagnetic-Spectrum. png (deleted). Transferred by Penubag on 15 May 2008. This file is published online under conditions imposed by a Creative Commons Attribution-Share Alike 2.5 Generic License; Armael, the copyright holder of the image on p. 124,

Index

Page numbers in **bold italics** refer to illustrations